The Hyperboloidal Foliation Method

Series in Applied and Computational Mathematics
ISSN: 2010-2739

Vol. 1: Analysis of Singularities for Partial Differential Equations
by Shuxing Chen (Fudan University, China)

Vol. 2: The Hyperboloidal Foliation Method
by Philippe G. LeFloch *(Université Pierre et Marie Curie, France)* &
Yue Ma *(Université Pierre et Marie Curie, France)*

Series in Applied and Computational Mathematics – Vol. 2

The Hyperboloidal Foliation Method

Philippe G. LeFloch
Yue Ma
Université Pierre et Marie Curie, France

World Scientific
NEW JERSEY • LONDON • SINGAPORE • BEIJING • SHANGHAI • HONG KONG • TAIPEI • CHENNAI

Published by

World Scientific Publishing Co. Pte. Ltd.
5 Toh Tuck Link, Singapore 596224
USA office: 27 Warren Street, Suite 401-402, Hackensack, NJ 07601
UK office: 57 Shelton Street, Covent Garden, London WC2H 9HE

Library of Congress Cataloging-in-Publication Data
LeFloch, Philippe G., 1962–
The hyperboloidal foliation method / Philippe G. LeFloch and Yue Ma, Universite Pierre et Marie Curie, France.
pages cm. -- (Series in applied and computational mathematics ; volume 2)
Includes bibliographical references and index.
ISBN 978-9814641623 (alk. paper)
1. Nonlinear wave equations. 2. Klein-Gordon equation. 3. Nonlinear systems. I. Ma, Yue (Mathematician) II. Title.
QA927.L3825 2015
530.1101'5169--dc23

2014039435

British Library Cataloguing-in-Publication Data
A catalogue record for this book is available from the British Library.

Printed in Singapore

Preface

The Hyperboloidal Foliation Method presented in this monograph is based on a $(3+1)$–foliation of Minkowski spacetime by hyperboloidal hypersurfaces. It allows us to establish global-in-time existence results for systems of nonlinear wave equations posed on a curved spacetime and to derive uniform energy bounds and optimal rates of decay in time. We are also able to encompass the wave equation and the Klein-Gordon equation in a unified framework and to establish a well-posedness theory for nonlinear wave-Klein-Gordon systems and a large class of nonlinear interactions.

The hyperboidal foliation of Minkowski spactime we rely upon in this book has the advantage of being *geometric in nature* and, especially, *invariant* under Lorentz transformations. As stated, our theory applies to many systems arising in mathematical physics and involving a massive scalar field, such as the Dirac-Klein-Gordon system. As it provides uniform energy bounds and optimal rates of decay in time, our method appears to be very robust and should extend to even more general systems.

We have built upon many earlier studies of nonlinear wave equations or Klein-Gordon equations, especially by Sergiu Klainerman, Demetri Christodoulou, Jalal Shatah, Alain Bachelot, and many others. The coupling of nonlinear wave-Klein-Gordon systems was first understood by Soichiro Katayama who succeeded to establish an existence theory of such systems.

Importantly, in developing the Hyperboloidal Foliation Method, we were inspired by earlier work on the Einstein equations of general relativity by Helmut Friedrich, Vincent Moncrief, and Anil Zenginoglu.

We are very grateful Soichiro Katayama for observations he made to the authors on a preliminary version of this monograph.

Last but not least, the authors are very grateful to their respective families for their strong support.

Philippe G. LeFloch and Yue Ma
Paris, September 2014
http://philippelefloch.org

Contents

Chapter 1

Introduction

1.1 Background and main objective

We are interested in nonlinear wave equations posed on the $(3 + 1)$-dimensional Minkowski spacetime and, especially, in models arising in mathematical physics and involving a nonlinear coupling with the Klein-Gordon equation. A typical example is provided by the Dirac-Klein-Gordon equation. (Cf. Section 1.5, below.) Our study is also motivated by the Einstein equations of general relativity when the matter model is a massive scalar field. The Klein-Gordon equation also describes nonlinear waves that propagate in fluids or in elastic materials.

The so-called 'vector field method' was introduced by Klainerman (1980, 1985, 1986, 1987). It is based on weighted norms defined from the conformal Killing fields of Minkowski spacetime and Sobolev-type arguments, and yields a global-in-time, well-posedness theory for the initial value problem for nonlinear wave equations, when the initial data have small amplitude and are "localized", that is, compactly supported.

This method relies on a bootstrap argument and on an analysis of the time decay of solutions. It applies to quadratic nonlinearities satisfying the so-called 'null condition' introduced by Klainerman and Christodoulou. A vast literature is now available on nonlinear wave equations. (Cf. Section 1.4 for further references.) As far as coupled systems of wave and Klein-Gordon equations are concerned, the current state-of-the-art is given by a recent work by Katayama (2012a) who succeded to extend the vector field method.

In this monograph, building upon these earlier works, we introduce a novel approach, which we refer to as the 'hyperboloidal foliation method' and we establish a global-in-time existence theory for a broad class of nonlinear wave-Klein-Gordon systems. In short, by working with suitably weigthed spacetime norms, we are able to cover the large class of quadratic nonlinearities, also recently treated by Katayama (2012a)), while our method demands limited regularity on the initial data and provides sharp bounds on the asymptotic profile of solutions. The hyperboloidal foliation method introduces a novel methodology, which takes its roots in an observation* by Hörmander (1997) for the Klein-Gordon equation.

Let us denote by $\Box$ the wave operator† in Minkowski space. For our purpose in this monograph, a simple (yet challenging) model of interest is provided by the following two equations which couple a wave equation with a Klein-Gordon equation

$$\begin{aligned} \Box u &= P(\partial u, \partial v), \\ \Box v + v &= Q(\partial u, \partial v), \end{aligned} \tag{1.1.1}$$

where the unknowns are the two scalar fields u, v. This model describes the nonlinear interactions between a massless scalar field and a massive one. Here, the nonlinear terms $P = P(\partial u, \partial v)$ and $Q = Q(\partial u, \partial v)$ are quadratic forms in the first-order spacetime derivatives $\partial u, \partial v$, and account for self-interactions as well as interactions between the two fields.

Recall that global-in-time existence results for nonlinear wave equations (without Klein-Gordon components) is established when the nonlinearities satisfy the null condition. (Cf. (1.2.4c), below.) On the other hand, the vector field method applies also to the nonlinear Klein-Gordon equation, as shown by Klainerman (1985). Recall also that the global existence problem for the nonlinear Klein-Gordon equation was also solved independently by Shatah (1985) with a different method.

However, when one attempts to tackle *coupled* systems of wave and Klein-Gordon equations like (1.1.1), one faces a major challenge due to the fact that one of the conformal Killing fields associated with the wave equation (the scaling vector field denoted below by $t\partial_t + r\partial_r$) is *not* a conformal Killing field for the Klein-Gordon equation and, therefore, can no longer be used in the vector field analysis. Katayama (2012a) succeeded to circumvent this difficulty and established a global existence theory for a class of coupled systems which includes (1.1.1). His method relies on a novel L^∞-L^∞ estimate. (See Section 1.4 for further historical background.)

---

*recalled in Section 2.1, below

†Our convention here is $\Box := \partial_t \partial_t - \sum_{a=1}^3 \partial_a \partial_a$.

1.2 Statement of the main result

The new method we provide in the present monograph relies on a fully geometric foliation. It is 'robust' as it is expected to be applicable to large classes of curved spacetimes and nonlinear hyperbolic equations.

We state here the main result that we will establish in this monograph. We are interested in the Cauchy problem for the following large class of **nonlinear systems of wave–Klein-Gordon equations**:

$$\begin{aligned}
&\Box w_i + G_i^{j\alpha\beta}(w,\partial w)\partial_\alpha\partial_\beta w_j + c_i^2 w_i = F_i(w,\partial w),\\
&w_i(B+1,x) = w_{i0},\\
&\partial_t w_i(B+1,x) = w_{i1},
\end{aligned} \tag{1.2.1}$$

in which the unknowns are the functions w_i ($1 \leqslant i \leqslant n_0$) defined on Minkowski space $\mathbb{R}^{3+1}$ and w_{i0}, w_{i1} are prescribed initial data. Throughout, Latin indices a, b, c will take values within $1, 2, 3$, while Greek indices α, β, γ take values within $0, 1, 2, 3$. Einstein's convention on repeated indices is in order throughout.

We assume the **symmetry conditions**

$$G_i^{j\alpha\beta} = G_j^{i\alpha\beta}, \quad G_i^{j\alpha\beta} = G_i^{j\beta\alpha} \tag{1.2.2}$$

and, for definiteness, the **wave-Klein-Gordon structure**

$$c_i \begin{cases} = 0, & 1 \leqslant i \leqslant j_0,\\ \geqslant \sigma, & j_0+1 \leqslant i \leqslant n_0, \end{cases} \tag{1.2.3}$$

where $\sigma > 0$ is a (constant, positive) lower bound for the mass coefficients of Klein-Gordon equations. We decompose the **"curved metric"** coefficients $G_i^{j\alpha\beta}(w,\partial w)$ and the **interaction terms** $F_i(w,\partial w)$ in the form

$$G_i^{j\alpha\beta}(w,\partial w) = A_i^{j\alpha\beta\gamma k}\partial_\gamma w_k + B_i^{j\alpha\beta k} w_k, \tag{1.2.4a}$$

$$F_i(w,\partial w) = P_i^{\alpha\beta jk}\partial_\alpha w_j \partial_\beta w_k + Q_i^{\alpha jk} w_k \partial_\alpha w_j + R_i^{jk} w_j w_k, \tag{1.2.4b}$$

in which we can restrict the summation in (1.2.4a) and (1.2.4b) to the range $j \leqslant k$. For simplicity in the presentation of the method and without genuine loss of generality, we focus on quadratic nonlinearities and assume that the coefficients $A_i^{j\alpha\beta\gamma k}$, $B_i^{j\alpha\beta k}$, $P_i^{\alpha\beta jk}$, $Q_i^{\alpha jk}$, and R_i^{jk} are constants.

In order to simplify the notation, we adopt the following index convention:

all indices i, j, k, l take the values $1, \dots, n_0$,

all indices $\hat{\imath}, \hat{\jmath}, \hat{k}, \hat{l}$ take the values $1, \dots, j_0$,

all indices $\check{\imath}, \check{\jmath}, \check{k}, \check{l}$ take the values $j_0+1, \dots, n_0$.

It is also convenient to write

$$u_{\widehat{\imath}} := w_{\widehat{\imath}}$$

for the components satisfying wave equations, or **wave components** for short, and, analogously,

$$v_{\check{\imath}} := w_{\check{\imath}}$$

for the components satisfying Klein-Gordon equations, or **Klein-Gordon components**.

Our main assumptions are the **null condition for wave components**

$$\begin{aligned} &A_{\widehat{\imath}}^{\widehat{\jmath}\alpha\beta\gamma\widehat{k}}\xi_\alpha\xi_\beta\xi_\gamma = B_{\widehat{\imath}}^{\widehat{\jmath}\alpha\beta\widehat{k}}\xi_\alpha\xi_\beta = P_{\widehat{\imath}}^{\alpha\beta\widehat{\jmath}\widehat{k}}\xi_\alpha\xi_\beta = 0 \\ &\text{whenever } (\xi_0)^2 - \sum_a (\xi_a)^2 = 0, \end{aligned} \tag{1.2.4c}$$

and the **non-blow-up condition**

$$B_i^{\check{\jmath}\alpha\beta\widehat{k}} = R_i^{j\widehat{k}} = R_i^{\widehat{\jmath}k} = 0. \tag{1.2.4d}$$

Moreover, we impose that

$$Q_i^{\alpha j\widehat{k}} = 0, \tag{1.2.4e}$$

which is our only genuine restriction* required for the present implementation of the hyperboloidal foliation method. We emphasize that

> the null condition in (1.2.4c) is imposed for the quadratic forms associated with the *wave components* only,

and that no such restriction is required for the Klein-Gordon components.

We observe† that (1.2.4d) combined with the symmetric condition (1.2.2) leads to the following restriction on $B_i^{j\alpha\beta k}$:

$$B_{\check{\jmath}}^{\widehat{\imath}\alpha\beta\widehat{k}} = 0. \tag{1.2.5}$$

This class of systems was also studied in Katayama (2012a) by a completely different approach.

We are now in a position to state the main result that we establish in the present monograph with the Hyperboloidal Foliation Method.

---

*When this condition is violated, solutions may not have the time decay and asymptotics of solutions to linear wave or Klein-Gordon equations in Minkowski space.

†as pointed out to the authors by S. Katayama

Theorem 1.2.1 (Global well-posedness theory). *Consider the initial value problem* (1.2.1) *with smooth initial data posed on the spacelike hypersurface* $\{t = B+1\}$ *of constant time and compactly supported in the ball* $\{t = B+1;\, |x| \leqslant B\}$. *Under the conditions* (1.2.2)–(1.2.4), *there exists a real* $\epsilon_0 > 0$ *such that, for all initial data* $w_{i0}, w_{i1} : \mathbb{R}^3 \to \mathbb{R}$ *satisfying the smallness condition*

$$\sum_i \|w_{i0}\|_{H^6(\mathbb{R}^3)} + \|w_{i1}\|_{H^5(\mathbb{R}^3)} < \epsilon_0, \tag{1.2.6}$$

the Cauchy problem (1.2.1) *admits a unique, smooth global-in-time solution. In addition, the energy of the wave components –that is,* $\sum_{|I|\leqslant 3} E_{m,c_i}(s, Z^I u_{\hat{\imath}})$ *defined in Section 2.1 below– remains globally bounded in time.*

In the special case $n_0 = j_0$, the system (1.2.1) contains only wave equations and the statement in Theorem 1.2.1 reduces to the classical existence result for quasilinear wave equations satisfying the null condition. Our method is somewhat simpler than the classical proof in this case, as we will show in Chapter 6. In the opposite direction, if we take $j_0 = 0$, the system under consideration contains Klein-Gordon components only, and our result reduces to the classical existence result for quasilinear Klein-Gordon equations.

An outline of this monograph is as follows. In Chapter 2, we introduce some basic notations on the hyperboloidal foliation and the associated energy, and we formulate our bootstrap assumptions. Chapter 3 is devoted to the derivation of fundamental properties of vector fields and commutators and their decompositions, which we will use throughout this book.

In Chapter 4, we discuss the null condition in the proposed semi-hyperboloidal frame and we derive preliminary estimates on first- and second-order derivatives of the solutions. In Chapter 5, we present some technical tools and, especially, a Sobolev inequality on hyperboloids and a Hardy inequality along the hyperboloidal foliation.

At this juncture, in Chapter 6 we apply our method to scalar semilinear wave equations and we provide a new proof of the standard existence result for such equations

We pursue our analysis, in Chapter 7, with fundamental estimates in the L^∞ and L^2 norms, which follow from our bootstrap assumptions. Chapter 8 is devoted to controlling certain 'second-order derivatives' of the wave components (in a sense explained therein), while Chapter 9 deals with quadratic terms satisfying the null condition and also discusses additional estimates that mainly rely on the time decay of solutions.

Next, in Chapter 10, we derive L^2 estimates for nonlinear interaction terms and, therefore, complete our bootstrap argument.

Finally, for the sake of completeness, in Chapter 11 we sketch the local-in-time existence theory. Furthermore, the bibliography at the end of this book provides the reader with further material of interest.

1.3 General strategy of proof

In this section, we present several key features of our method, while referring to Chapter 2 for the relevant notions (hyperboloidal foliation, bootstrap estimates, etc.).

- **Hyperboloidal foliation.**
  Most importantly, in this book we propose to work with the *family of hyperboloids* which generate a foliation of the interior of the light cone in Minkowski spacetime. (Cf. (2.1.4), below.) In contrast with other foliations of Minkowski space which are adopted in the literature, the hyperboidal foliation has the advantage of being *geometric in nature* and *invariant* under Lorentz transformations. It is therefore quite natural to search for an estimate of the energy defined on these hypersurfaces, rather than the energy defined on flat hypersurfaces of constant time, as is classically done.
- **The semi-hyperboloidal frame.**
  Furthermore, to the hyperboloidal foliation we attach a *semi-hyperboloidal frame* (as we call it), which consists of three vectors tangent to the hyperboids plus a timelike vector. This frame has several advantages in the analysis in comparison with, for instance, the 'null frame' which is often used in the literature and, instead, is defined from vectors tangent to the light cone. Importantly, the semi-hyperboloidal frame is *regular* (in the interior of the light cone, which is the region of interest), while the null frame is singular at the center ($\{r = 0\}$, say).
- **Decomposition of the wave operator.**
  We also introduce a decomposition of the wave operator $\Box$ with respect to the semi-hyperboloidal frame, which yields us an expression of the second-order time derivative $\partial_t\partial_t$ of the wave compo-

nents in terms of better-behaved derivatives. (Cf. Proposition 2.2.1, below.)

- **The hyperboloidal energy.**
Our method takes advantage of the *full* expression* of the energy flux induced on the hyperboloids in order to estimate certain *weigthed derivatives on the hyperboloids.* This appears to be essential in order to encompass wave equations and Klein-Gordon equations in a single framework.
- **Sobolev inequality on hyperboloids.**
In order to establish decay estimates (or L^∞ estimates) on the solutions, we must control various commutators of fields and of operators and, next, apply suitable embedding theorems. To this purpose, we rely on a Sobolev inequality on hyperboloids, first derived by Hörmander (1997).
- **Hardy inequality on hyperboloids.**
Furthermore, we also need a new embedding estimate, that is, a Hardy inequality on hyperboloids, which we establish in this book and is essential in eventually deriving an L^2 estimate on the 'metric coefficients'. Section 5.3 for a precise statement.
- **Bootstrap strategy and hierarchy of energy bounds**
Our bootstrap formulation below consists of a *hierarchy of energy bounds*, involving several levels of regularity of the wave components and the Klein-Gordon components. This rather involved bootstrap argument is necessary (and natural) in order to handle the coupling of wave equations and Klein-Gordon equations: the derivatives of different order of Klein-Gordon components enjoy different decay behaviors and different energy bounds.

We refer to Chapter 2 for further details and continue with several observations concerning the scope of Theorem 1.2.1.

1. As stated in the theorem, the energy of the wave components, that is, the quantity $E_{m,c_i}(s, Z^I u_{\hat{\imath}})$ (defined in Section 2.1) remains globally bounded for all $|I| \leqslant 3$, that is, up to fourth-order derivatives. Hence, the wave components have not only 'small' amplitude but also 'small' energy. At the end of Chapter 2, we also establish that the standard flat energy (that is, the quantity $\|\partial_\alpha Z^I u_{\hat{\imath}}(t, \cdot)\|_{L^2(\mathbb{R}^3)}$ defined at the end of Chapter 2, below) is also uniformly bounded for all times. Standard methods of proof

*In constrast, Hörmander (1997) only sought for a control of the zero-order term of the energy.

lead to possibly *unbounded* high-order energies.

2. We also emphasize that, in Theorem 1.2.1, the initial data belong to H^6, while Katayama (2012a) assumes a very high regularity on the initial data (that is, a bound on the first 19 derivatives). In this latter paper however, the initial data need not be compactly supported, but have sufficient decay in all spatial directions.

3. In principle, H^4 would be the optimal regularity in order to work with a uniformly bounded metric and to apply the vector field technique: namely, to guarantee the coercivity of, both, the flat and the hyperboloidal energy functionals, we need a sup-norm bound of the second-order derivatives the 'curved metric' terms $G_i^{j\alpha\beta}(w, \partial w)$. In spatial dimension three, Sobolev's embedding theorem $H^m \subset L^\infty$ holds, provided $m > 3/2$. Allowing only integer exponents, we see that H^4 would be optimal.

4. Certain nonlinear interaction terms may lead to a *finite time blow-up* of the solutions, especially

$$u_{\hat{\imath}} u_{\hat{\jmath}}, \quad u_{\hat{\imath}} \partial_\alpha u_{\hat{\jmath}}, \quad u_{\hat{\imath}} v_{\check{\jmath}}, \quad u_{\hat{\imath}} \partial_\alpha v_{\check{\jmath}}$$

and are, therefore, naturally excluded in Theorem 1.2.1.

5. In short, by denoting by Q an arbitrary quadratic nonlinearity and by N an arbitrary quadratic null form and by using the notation u, v for arbitrary wave/Klein-Gordon components, the terms allowed in Theorem 1.2.1 for wave equations are

$$\begin{aligned}
&Q(v,v), \quad Q(v,\partial v), \quad Q(v,\partial u), \quad Q(v,\partial\partial v) \quad Q(v,\partial\partial u),\\
&Q(\partial v,\partial v), \quad Q(\partial v,\partial u), \quad Q(\partial v,\partial\partial v), \quad Q(\partial v,\partial\partial u),\\
&Q(\partial u,\partial\partial v),\\
&N(u,\partial\partial u), \quad N(\partial u,\partial u), \quad N(\partial u,\partial\partial u),
\end{aligned}$$

while, in Klein-Gordon equations, we can include the terms

$$\begin{aligned}
&Q(v,v), \quad Q(v,\partial v), \quad Q(v,\partial u), \quad Q(v,\partial\partial v), \quad Q(v,\partial\partial u),\\
&Q(\partial v,\partial v), \quad Q(\partial v,\partial u), \quad Q(\partial v,\partial\partial v), \quad Q(\partial v,\partial\partial u),\\
&Q(\partial u,\partial\partial v), \quad Q(\partial u,\partial u), \quad Q(\partial u,\partial\partial u).
\end{aligned}$$

1.4 Further references on earlier works

For a background on the vector field method and the global well-posedness for nonlinear wave equations, in additional to the references cited earlier, especially the pioneering paper Klainerman (1980), we refer to the textbooks Hörmander (1997) and Sogge (2008). Additional background on nonlinear wave equations is found in Strauss (1989). We do not attempt to review the large literature and only mention some selected results, while referring the reader to the bibliography at the end of this monograph.

As already mentioned, the first results of global existence for nonlinear wave equations in three spatial dimensions were established by Klainerman (1986) and Christodoulou (1986) under the assumption that the nonlinearities satisfy the null condition and when the equation is posed in the flat Minkowski space. A (very) large literature is available for equations posed on curved spaces and, once more, we do not try to be exhaustive. We refer to Lindblad (1990), Klainerman and Sideris (1996), Klainerman and Selberg (1997), Klainerman and Machedon (1997), Bahouri and Chemin (1999), Tataru (2000, 2001, 2002), Alinhac (2004, 2006), Lindblad and Rodnianski (2005), and Lindblad, Nakamura, and Sogge (2012).

The Klein-Gordon equation on curved spaces was also studied in Bachelot (1994, 2011).

Since the decay of solutions to the (linear) Klein-Gordon equation is $t^{-d/2}$ in dimension $d \geqslant 1$, the decay function $t^{-d/2}$ is not integrable in dimension two and specific arguments are required in two dimensions: Delort, Fang, and Xue (2004) have treated quadratic quasilinear Klein-Gordon systems in two space dimensions and, more precisely, coupled systems of two equations with masses satisfying $m_1 \neq 2m_2$ and $m_2 \neq 2m_1$ with general nonlinearities. Furthermore, Delort, Fang, and Xue (2004) could treat the case of equality when the null condition is assumed. This work simplified and generalized (by including resonant cases) the earlier works by Ozawa, Tsutaya and Tsutsumi (1995, 1996), Tsutsumi (2003a,b) and Sunagawa (2003, 2004). See also Katayama, Ozawa, and Sunagawa (2012) for the algebraic characterization of the null condition and the asymptotic behavior of solutions, as well as Kawahara and Sunagawa (2011) for a condition weaker than the null condition.

More recently, Germain (2010) revisited the global existence theory in dimension three for coupled Klein-Gordon equations with different speeds and systematically analyzed resonance effects. Systems of wave equations for different speeds were also studied by Yokohama (2000) and Sideris and Tu (2001) under the null condition. See also Hoshiga and Kubo (2000), Katayama (2013), Katayama and Yokoyama (2006), and Kubota and Yokoyama (2001).

On the other hand, Klein-Gordon equations in one space dimension are treated by quite different methods. (Cf. Delort (2001); Sunagawa (2003); Candy (2013).)

The problem of global existence for coupled systems of wave and Klein-Gordon equations have attracted much less attention so far in the literature. In addition to the references already quoted, let us mention Bachelot (1988) who first treated the Dirac-Klein-Gordon system. Furthermore, results on the blow-up of solutions were established by John (1979, 1981) and, more recently, Alinhac (2000).

More recenty, a novel method to study nonlinear wave equations (applicable also to other dispersive systems) was introduced in Shatah (2010), Germain, Masmoudi, and Shatah (2012), Pusateri and Shatah (2013), Pusateri (2013), and Bernicot and Germain (2014), which is based on an analysis of space-time resonances. See also the review by Lannes (2013).

Hyperboloidal foliations were used first by Friedrich (1981, 1983, 2002) in order to establish a global existence result for the Einstein equations. His proof was based on a conformal transformation of the Einstein equations and an analysis of the regularity of its solutions at infinity. This was motivated by earlier work by Penrose (1963) on the compactification of spacetimes. This idea was later developed by Frauendiener (1998, 2002, 2004) and Rinne and Moncrief (2013). The importance of hyperboloidal foliations for general hyperbolic systems was emphasized in Zenginoglu (2008, 2011) in order to numerically compute solutions within an unbounded domain. The general standpoint in these works is that, by compactification of the spacetime, one can conveniently formulate an 'artificial' outer boundary and numerically compute asymptotic properties of interest.

1.5 Examples and applications

The Maxwell-Klein-Gordon system

The theory presented in this book applies to many systems arising in mathematical physics, and we present here a few of them. For instance, the Maxwell-Klein-Gordon system in Coulomb gauge takes the form of a system of nonlinear wave equations for real-valued unknown A_j and a complex-valued field ϕ

$$\begin{aligned} -\Box A_j &= -\Im(\phi\overline{\partial_j\phi}) + |\phi|^2 A_j - \partial_j\partial_t A_0, \\ -\Box\phi &= 2\sqrt{-1}\big(- A^j\partial_j\phi + A_0\partial_t\phi\big) + \sqrt{-1}\partial_t A_0\phi + \big(A_\alpha A^\alpha + m^2\big)\phi, \end{aligned} \tag{1.5.1}$$

with auxillary unknown A_0 given by

$$\Delta A_0 = -\Im(\phi\overline{\partial_t\phi}) + |\phi|^2 A_0, \tag{1.5.2}$$

supplemented with an elliptic constraint equation imposed on the initial data

$$\partial^j A_j = 0. \tag{1.5.3}$$

The Dirac-Klein-Gordon system

Consider next the following coupling between the Dirac equation and the wave or Klein-Gordon equation (with $m, \sigma \geqslant 0$):

$$\begin{aligned} & -\sqrt{-1} \sum_{\alpha=0}^{3} \Gamma_\alpha \, \partial_\alpha \psi + m\psi = \beta \, v \Gamma_0 \Gamma_1 \Gamma_2 \Gamma_3 \psi, \\ & \Box v + \sigma^2 v = \psi^\dagger K \psi, \end{aligned} \tag{1.5.4}$$

in which the unknown are the ($\mathbb{C}^4$-valued) spinor field ψ and the (real-valued) scalar field v. We have denoted by $\psi^\dagger$ the complex conjugate transpose of ψ. Here, β is a coupling constant and K a constant 4×4 matrix, while the 4×4 matrices Γ_α are the so-called Dirac matrices which are essentially characterized by the commutation conditions

$$\Gamma_\alpha \Gamma_\beta + \Gamma_\beta \Gamma_\alpha = -2I \, m_{\alpha\beta}, \tag{1.5.5}$$

where $m_{\alpha\beta}$ is the Minkowski metric $\operatorname{diag}(1, -1, -1, -1)$ and I denotes the 4×4 identity matrix. From the Dirac equation, on can deduce second-order equations for real-valued unknowns (so that our theory applies): this is done by composing the Dirac operator with itself (since, roughly speaking, the Dirac operator is the "square-root" of the wave operator) and then considering the real and imaginary parts of ψ.

The Einstein equations

Although our theory in its present form does not directly apply to the Einstein equations of general relativity

$$G_{\alpha\beta} = T_{\alpha\beta}, \tag{1.5.6}$$

it is nonetheless motivated by this system and we expect a suitable extension of our method to apply to (1.5.6). The left-hand side $G_{\alpha\beta}$ of (1.5.6) is the Einstein tensor of a spacetime (M, g), that is, a Lorentzian $(3+1)$-dimensional manifold, while the right-hand side $T_{\alpha\beta}$ denotes the energy momentum tensor of a matter field, which in our context can be assumed to be a set of massless and massive scalar fields.

Chapter 2

The hyperboloidal foliation and the bootstrap strategy

2.1 The hyperboloidal foliation and the Lorentz boosts

We will work with the foliation of the interior of the light cone in Minkowski spacetime $\mathbb{R}^{3+1}$, defined as below.

We introduce the **hyperboloidal hypersurfaces**

$$\mathcal{H}_s := \big\{(t,x)\,/\,t>0;\, t^2-|x|^2 = s^2\big\} \tag{2.1.1}$$

with hyperbolic radius $s>0$, where $(t,x)=(t,x^a)=(t,x^1,x^2,x^3)$ denote Cartesian coordinates, and we write $r^2 := |x|^2 = \sum_a (x^a)^2$. We then consider the interior of the (future) light-cone

$$\mathcal{K} := \big\{(t,x)\,/\,|x|<t-1\big\} \tag{2.1.2}$$

and, with $s_1 > s_0 > 1$, as well as the truncated conical region

$$\begin{aligned}\mathcal{K}_{[s_0,s_1]} :=& \big\{(t,x)\,/\,|x|<t-1, \quad (s_0)^2 \leqslant t^2-|x|^2 \leqslant (s_1)^2, t>0\big\}\\ =& \bigcup_{s_0\leqslant s\leqslant s_1} (\mathcal{H}_s\cap\mathcal{K}).\end{aligned} \tag{2.1.3}$$

This set is thus limited by two hyperboloids and is naturally foliated by hyperboloids. See Fig. 2.1 for a display of the set $\mathcal{K}_{[s_0,s_1]}$.

Taking now $s_1 = +\infty$, we will use the notation

$$\begin{aligned}\mathcal{K}_{[s_0,+\infty)} :=& \big\{(t,x)\,/\,|x|<t-1,\, (s_0)^2 \leqslant t^2-|x|^2\big\}\\ =& \bigcup_{s\geqslant s_0} (\mathcal{H}_s\cap\mathcal{K}).\end{aligned} \tag{2.1.4}$$

We refer to Fig. 2.2 for a display of this set.

In the following, we will be interested in functions supported in the conical region $\mathcal{K}_{[s_0,+\infty)}$. Observe that the set $\mathcal{K}_{[s_0,+\infty)}$ is neither closed nor open and a function supported in this set, by definition, vanishes near the future light cone $\{r=t-1\}$ but can be non-vanishing on the surface $\mathcal{H}_{s_0}$.

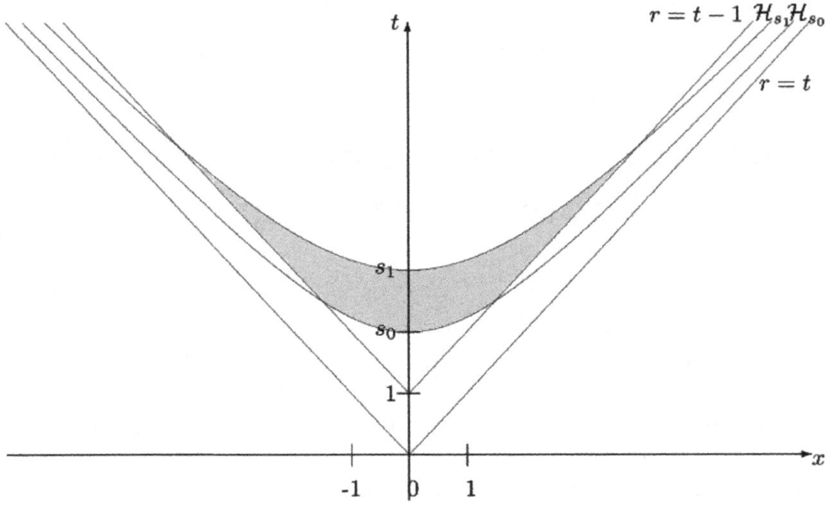

Fig. 2.1 The set $\mathcal{K}_{[s_0,s_1]}$.

The region $\mathcal{K} \cap \{|x| \leqslant t/2\}$ will be also of interest in our estimates below, when we will investigate the behavior of solutions away from the light cone. Note in passing that the uniform estimate $t \leqslant \frac{2\sqrt{3}}{3} s$ holds in $\mathcal{K} \cap \{|x| \leqslant t/2\}$.

We consider first the Klein-Gordon equation

$$\begin{aligned} &\Box u + \sigma^2 u = f, \\ &u(t,x)|_{\mathcal{H}_{s_0}} = u_0(t,x), \quad u_t(t,x)|_{\mathcal{H}_{s_0}} = u_1(t,x), \end{aligned} \tag{2.1.5}$$

with given $\sigma, s_0 = B + 1 > 1$. Recall that the symbol $\Box$ denotes the wave operator in Minkowski spacetime whose metric has the signature $(1,-1,-1,-1)$. In (2.1.5), the initial data u_0, u_1 are prescribed and compactly supported in the ball $\big\{|x| \leqslant B\big\}$ of radius $B = s_0 - 1$. The source-term function f is supported in $\mathcal{K}_{[s_0,+\infty)}$, so that, by the principle of propagation at finite speed, the solution $u = u(t,x)$ to (2.1.5) is also supported in $\mathcal{K}_{[s_0,+\infty)}$. In the same manner, the solution of the main system (1.2.1) studied in this monograph is also supported in $\mathcal{K}_{[s_0,+\infty)}$ and vanishes in a neighborhood of the light cone $\big\{r = t - 1\big\}$.

Next, let us introduce the **hyperbolic rotations** or **Lorentz boosts** (by rising and lowering the indices with the Minkowski metric with signa-

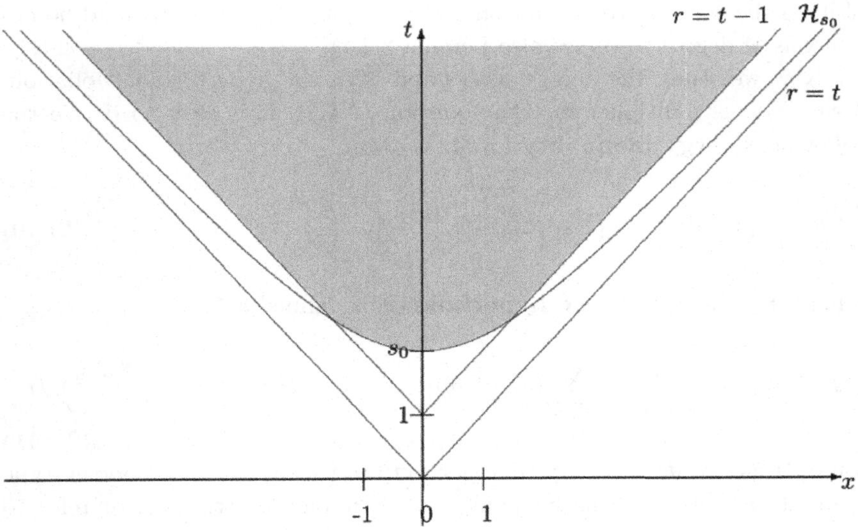

Fig. 2.2 The set $\mathcal{K}_{[s_0,+\infty)}$.

ture $(+,-,-,-)$)

$$L_a := -x_a \partial_0 + x_0 \partial_a = x^a \partial_t + t\partial_a, \tag{2.1.6}$$

which are tangent vectors to the hyperboloids. Denote by $\mathscr{Z}$ the family of **admissible vector fields** consisting of all vectors

$$Z_\alpha := \partial_\alpha, \qquad Z_{3+a} := L_a. \tag{2.1.7}$$

Observe that, for any Z, $Z' \in \mathscr{Z}$, the Lie bracket $[Z, Z']$ also belongs to $\mathscr{Z}$, so that this set is a Lie algebra. For any multi-index $I = (\alpha_1, \alpha_2, \dots, \alpha_m)$ of length $|I| := m$, we denote by Z^I the m-th order differential operator $Z^I := Z_{\alpha_1} \dots Z_{\alpha_m}$. We also denote by ∂^I the m-th order derivative operator $\partial^I := \partial_{\alpha_1}\partial_{\alpha_2} \dots \partial_{\alpha_m}$ (here $0 \leqslant \alpha_i \leqslant 3$) and L^I the m-th order derivative operator $L^I := L_{\alpha_1} L_{\alpha_2} \dots L_{\alpha_m}$ (here $4 \leqslant \alpha_j \leqslant 6$).

Since we will be working within $\mathcal{K}$, we have $|x^a/t| \leqslant 1$ in $\mathcal{K}$ and, therefore, the **spatial rotations**

$$\Omega_{ab} := x^a \partial_b - x^b \partial_a \tag{2.1.8}$$

need not be included explicitly in our analysis, since these fields can be recovered from $\mathscr{Z}$ via the identities

$$\Omega_{ab} = \frac{x^a}{t} L_b - \frac{x^b}{t} L_a. \tag{2.1.9}$$

Within the cone $\mathcal{K}$, the coefficients x^a/t are smooth, bounded, and homogeneous of degree zero (see also Lemma 2.2.1).

Now we study the energy associated with the hyperboloidal foliation. Using $\partial_t u$ as multiplier for the equation (2.1.5), it is easy to derive the following **energy inequality** for all $s_1 \geqslant s_0$:

$$\big(E_{m,\sigma}(s_1,u)\big)^{1/2} \leqslant \big(E_{m,\sigma}(s_0,u)\big)^{1/2} + \int_{s_0}^{s_1}\bigg(\int_{\mathcal{H}_s} f^2\,dx\bigg)^{1/2} ds, \tag{2.1.10}$$

where the **energy on the hyperboloids** is defined as*

$$E_{m,\sigma}(s_1,u) := \int_{\mathcal{H}_{s_1}} \Big(\sum_{a=1}^{3}\big((x^a/t)\partial_t u + \partial_a u\big)^2 + ((s_1/t)\partial_t u)^2 + \sigma^2 u^2\Big)\,dx \tag{2.1.11}$$

with $dx = dx^1dx^2dx^3$. Hörmander (1997)) established a Sobolev-type estimate adapted to this inequality (cf. Lemma 7.6.1 therein, or refer to (5.1.1) in Chapter 5, below) and arrived at the L^∞ estimate

$$\begin{aligned} &\sup_{\mathcal{H}_{s_1}} t^{3/2}|u| \\ &\leqslant C\sum_{|I|\leqslant 2} E_{m,\sigma}(s_1,Z^I u)^{1/2} \\ &\leqslant C\sum_{|I|\leqslant 2} E_{m,\sigma}(s_0,Z^I u)^{1/2} + C\sum_{|I|\leqslant 2}\int_{s_0}^{s_1}\bigg(\int_{\mathcal{H}_s}(Z^I f)^2\,dx\bigg)^{1/2} ds, \end{aligned} \tag{2.1.12}$$

where the summations are over all admissible vector fields in $\mathscr{Z}$. Importantly, this argument yields the (optimal) rate of decay $t^{-3/2}$ enjoyed by solutions to the Klein-Gordon equation.

2.2 Semi-hyperboloidal frame

Our analysis in the present work is based on the **semi-hyperboloidal frame** —as we call it— defined by†

$$\underline{\partial}_0 := \partial_t, \qquad \underline{\partial}_a := t^{-1}L_a = \big(x^a/t\big)\partial_t + \partial_a. \tag{2.2.1}$$

*The subscript refers to the Minkowski metric.

†In contrast, a standard method relies on the so-called null frame containing two null vectors tangent to the light cone.

The transition matrices between the semi-hyperboloidal frame and the natural frame ∂_β, that is, $\underline{\partial}_\alpha = \Phi_\alpha^\beta \partial_\beta$ and $\partial_\alpha = \Psi_\alpha^\beta \underline{\partial}_\beta$ are found to be

$$\Phi := \begin{pmatrix} 1 & 0\,0\,0 \\ x^1/t & 1\,0\,0 \\ x^2/t & 0\,1\,0 \\ x^3/t & 0\,0\,1 \end{pmatrix}, \qquad \Psi := \Phi^{-1} = \begin{pmatrix} 1 & 0\,0\,0 \\ -x^1/t & 1\,0\,0 \\ -x^2/t & 0\,1\,0 \\ -x^3/t & 0\,0\,1 \end{pmatrix}. \tag{2.2.2}$$

With our choice of frame, the matrices Φ, Ψ are smooth within the cone $\mathcal{K}$.

We adopt the following notation and convention. In order to express the components of a tensor in a frame, we always use Roman font with upper and lower indices for its components in the natural frame, while we use *underlined* Roman font for its components in the semi-hyperboloidal frame. Hence, a tensor is expressed as $T = T^{\alpha\beta}\partial_\alpha \otimes \partial_\beta$ in natural frame, and as $T = \underline{T}^{\alpha\beta}\underline{\partial}_\alpha \otimes \underline{\partial}_\beta$ in the semi-hyperboloidal frame. For example, the Minkowski metric is expressed in the semi-hyperboloidal frame as

$$m = \underline{m}^{\alpha\beta}\underline{\partial}_\alpha \otimes \underline{\partial}_\beta$$

with

$$\left(\underline{m}^{\alpha\beta}\right) = \begin{pmatrix} s^2/t^2 & x^1/t & x^2/t & x^3/t \\ x^1/t & -1 & 0 & 0 \\ x^2/t & 0 & -1 & 0 \\ x^3/t & 0 & 0 & -1 \end{pmatrix} \tag{2.2.3}$$

and

$$\left(\underline{m}_{\alpha\beta}\right) = \begin{pmatrix} 1 & x^1/t & x^2/t & x^3/t \\ x^1/t & (x^1/t)^2 - 1 & x^1x^2/t^2 & x^1x^3/t^2 \\ x^2/t & x^2x^1/t^2 & (x^2/t)^2 - 1 & x^2x^3/t^2 \\ x^3/t & x^3x^1/t^2 & x^3x^2/t^2 & (x^3/t)^2 - 1 \end{pmatrix}. \tag{2.2.4}$$

Similarly, a second-order differential operator $T^{\alpha\beta}\partial_\alpha\partial_\beta$ can be written in the semi-hyperboloidal frame, so that by writing $T^{\alpha\beta} = \underline{T}^{\alpha'\beta'}\Phi_{\beta'}^{\beta}\Phi_{\alpha'}^{\alpha}$ (with obvious notation), we obtain the following decomposition formula for any function u

$$T^{\alpha\beta}\partial_\alpha\partial_\beta u = \underline{T}^{\alpha\beta}\underline{\partial}_\alpha\underline{\partial}_\beta u + T^{\alpha\beta}(\partial_\alpha\Psi_\beta^{\beta'})\underline{\partial}_{\beta'}u. \tag{2.2.5}$$

In particular, for the wave operator, we obtain

$$\Box u = \underline{m}^{\alpha\beta}\underline{\partial}_\alpha\underline{\partial}_\beta u + m^{\alpha\beta}\left(\partial_\alpha\Psi_{\beta'}^{\beta}\right)\underline{\partial}_{\beta'}u, \tag{2.2.6}$$

where we know that $\underline{m}^{00} = (t^2 - r^2)/t^2 = s^2/t^2$. This leads us immediately to the following key identity.

Proposition 2.2.1 (Semi-hyperboloidal decomposition of $\Box$). *In the semi-hyperboloidal frame, the wave operator admits the decomposition*

$$(s/t)^2\underline{\partial}_0\underline{\partial}_0 u = \Box u - \underline{m}^{0a}\underline{\partial}_0\underline{\partial}_a u - \underline{m}^{a0}\underline{\partial}_a\underline{\partial}_0 u - \underline{m}^{ab}\underline{\partial}_a\underline{\partial}_b u \\ - m^{\alpha\beta}\big(\partial_\alpha \Psi_{\beta'}^{\beta}\big)\underline{\partial}_{\beta'} u. \tag{2.2.7}$$

We will also need the following property.

Proposition 2.2.2. *For any two-tensor T defined in the cone $\mathcal{K}$ and for all indices α, β, I and admissible field Z, one has*

$$\big|Z^I \underline{T}^{\alpha\beta}\big| \lesssim \sum_{\substack{\alpha',\beta' \\ |I'|\leqslant|I|}} \big|Z^{I'} T^{\alpha'\beta'}\big| \qquad \text{in } \mathcal{K}.$$

The proof of this result (given below) will rely on the following two lemmas.

Lemma 2.2.1 (Homogeneity lemma). *Let $f = f(t,x)$ be a smooth function defined in the closed region $\{t \geqslant 1, |x| \leqslant t\}$ and assumed to be homogeneous of degree η in the sense that:*

$$f(pt, px) = p^\eta f(t,x) \quad \text{for all } p \geqslant 1/t. \tag{2.2.8}$$

For all multi-indices I_1, I_2, the following estimate holds for some positive constant $C(n, |I_1|, |I_2|, f)$:

$$\big|\partial^{I_1} Z^{I_2} f(t,x)\big| \leqslant C(n, |I_1|, |I_2|, f)\, t^{-|I_1|+\eta} \qquad \text{in } \mathcal{K} = \big\{|x| < t-1\big\}. \tag{2.2.9}$$

Proof. We observe that $\partial_a f$ is homogeneous of degree $\eta - 1$ and $L_a f$ is homogeneous of degree η. We also observe that if some functions f_i are homogeneous of degree η_i, then the product $\prod_i f_i$ is homogeneous of degree $\sum_i \eta_i$ and then any $Z^I f$ is homogeneous.

We claim that the degree of homogeneity of $Z^I f$, denoted by η', is not higher than η. This can be checked by induction, as follows. Namely, this is clear when $|I| = 1$. Moreover, assume that for all $|I| \leqslant m$, $Z^I f$ is homogeneous of degree $\eta' \leqslant \eta$, then we now check the same property for all $|I| = m+1$. Namely, assume that $Z^I = Z_1 Z^{I'}$ with $|I'| = m$. When $Z_1 = \partial_\alpha$, then $Z^I f = \partial_\alpha\big(Z^{I'} f\big)$ and we observe that $Z^{I'} f$ is homogeneous of degree $\eta' \leqslant \eta$. We see that $\partial_\alpha\big(Z^{I'} f\big)$ is again homogeneous of degree $\eta' - 1 < \eta' \leqslant \eta$. When $Z_1 = L_a$, then $L_a\big(Z^{I'} f\big)$ is homogeneous of degree $\eta' \leqslant \eta$. This completes the induction argument that $Z^I f$ is homogeneous of degree η at most.

Furthermore, observe that if f is homogeneous of degree η, then $\partial^{I_1} f$ is homogeneous of degree $\eta - |I_1|$. This is so since $\partial_\alpha f$ is homogeneous of degree $\eta - 1$. Hence, $\partial^{I_1} Z^{I_2} f$ is homogeneous of degree $\eta - |I_1|$ and is smooth within the cone $\mathcal{K}$.

Next, in order to establish (2.2.9), we take $(t, x) \in \mathcal{K}$ and compute

$$\partial^{I_1} Z^{I_2} f(t,x) = t^{-|I_1|+\eta'} \partial^{I_1} Z^{I_2} f(1, x/t) \qquad \text{with } \eta' \leqslant \eta. \tag{2.2.10}$$

By a continuity argument in the compact set $\{t = 1, |x| \leqslant 1\}$, there exists a positive constant $C(n, |I_1|, |I_2|)$ such that

$$\sup_{|x| \leqslant 1} \left|\partial^{I_1} Z^{I_2} f(1,x)\right| \leqslant C(n, |I_1|, |I_2|),$$

so that, in view of (2.2.10), the desired result is proven. □

We now estimate the coefficients of the matrices Φ and Ψ. First, we observe that x^a/t is homogeneous of degree 0 and smooth in the closed region $\{t \geqslant 1, |x| \leqslant t\}$.

Lemma 2.2.2 (Changes of frame). *With the notation above, the following two estimates hold for all multi-indices I_1, I_2:*

$$\left|\partial^{I_1} Z^{I_2} \Phi_\alpha^\beta\right| \leqslant C(n, |I_1|, |I_2|)\, t^{-|I_1|} \qquad \text{in } \mathcal{K}. \tag{2.2.11}$$

$$\left|\partial^{I_1} Z^{I_2} \Psi_\alpha^\beta\right| \leqslant C(n, |I_1|, |I_2|)\, t^{-|I_1|} \qquad \text{in } \mathcal{K}. \tag{2.2.12}$$

Proof of Proposition 2.2.2. From $\underline{T}^{\alpha\beta} = T^{\alpha'\beta'} \Psi_{\alpha'}^\alpha \Psi_{\beta'}^\alpha$ we find

$$Z^I \underline{T}^{\alpha\beta} = Z^I \left(T^{\alpha'\beta'} \Psi_{\alpha'}^\alpha \Psi_{\beta'}^\alpha\right) = \sum_{I_1 + I_2 = I} Z^{I_1}\left(\Psi_{\alpha'}^\alpha \Psi_{\beta'}^\alpha\right) Z^{I_2} T^{\alpha'\beta'}$$

and, therefore,

$$\begin{aligned}\left|Z^I \underline{T}^{\alpha\beta}\right| &\leqslant \sum_{I_1 + I_2 = I} \left|Z^{I_1}\left(\Psi_{\alpha'}^\alpha \Psi_{\beta'}^\alpha\right)\right| \left|Z^{I_2} T^{\alpha'\beta'}\right| \\ &\leqslant C(|I|) \sum_{|I_2| \leqslant |I|} \left|Z^{I_2} T^{\alpha'\beta'}\right|.\end{aligned}$$

Here, we have used that $\left|Z^{I_1}\left(\Psi_{\alpha'}^\alpha \Psi_{\beta'}^\alpha\right)\right| \leqslant C(|I_1|)$ which follows from (2.2.12). □

2.3 Energy estimate for the hyperboloidal foliation

As presented in Chapter 1, we are interested in the following class of wave-Klein-Gordon type systems

$$\begin{aligned} &\Box w_i + G_i^{j\alpha\beta}\partial_\alpha\partial_\beta w_j + c_i^2 w_i = F_i, \\ &w_i|_{\mathcal{H}_{s_0}} = w_{i0}, \quad \partial_t w_i|_{\mathcal{H}_{s_0}} = w_{i1}, \end{aligned} \tag{2.3.1}$$

with unknowns w_i ($1 \leqslant i \leqslant n_0$), where the metric $G_i^{j\alpha\beta}$ (with some abuse of notation) and the source-terms F_i are supported in the cone $\mathcal{K}_{[s_0,+\infty)}$, and w_{i0}, w_{i1} are supported on the initial hypersurface $\mathcal{H}_{s_0} \cap \mathcal{K}$. To guarantee the hyperbolicity, we assume the symmetry conditions (1.2.2). For definiteness, we may also assume (1.2.3) on the constants c_i.

We introduce the following energy associated with the Minkowski metric on each hyperboloid $\mathcal{H}_s$:

$$\begin{aligned} &E_{m,c_i}(s, w_i) \\ &:= \int_{\mathcal{H}_s} \Big((\partial_t w_i)^2 + \sum_a (\partial_a w_i)^2 + (2x^a/t)\partial_t w_i \partial_a w_i + c_i^2 w_i^2 \Big)\, dx \\ &= \int_{\mathcal{H}_s} \Big(\sum_a (\underline{\partial}_a w_i)^2 + \big((s/t)\partial_t w_i\big)^2 + c_i^2 w_i^2 \Big)\, dx \\ &= \int_{\mathcal{H}_s} \Big(\sum_a \big((s/t)\partial_a w_i\big)^2 + t^{-2}(S w_i)^2 + t^{-2} \sum_{a<b} \big(\Omega_{ab} w_i\big)^2 + c_i^2 w_i^2 \Big)\, dx, \end{aligned} \tag{2.3.2}$$

where we use the notation (2.1.8) and

$$t^{-1} S := \partial_t + \sum_a (x^a/t)\partial_a. \tag{2.3.3}$$

When $c_i = 0$, we may also write $E_m(s, w_i) := E_{m,0}(s, w_i)$ for short.

On the other hand, the curved energy which is naturally associated with the principal part of (2.3.1) is defined as

$$\begin{aligned} E_{G,c_i}(s, w_i) := E_{m,c_i}(s, w_i) &+ 2\int_{\mathcal{H}_s} \big(\partial_t w_i \partial_\beta w_j G_i^{j\alpha\beta}\big)_{0\leqslant\alpha\leqslant 3} \cdot (1, -x/t)\, dx \\ &- \int_{\mathcal{H}_s} \big(\partial_\alpha w_i \partial_\beta w_j G_i^{j\alpha\beta}\big)\, dx, \end{aligned} \tag{2.3.4}$$

where the second term in the right-hand side involves the Euclidian inner product of the vectors $\big(\partial_t w_i \partial_\beta w_j G_i^{j\alpha\beta}\big)_{0\leqslant\alpha\leqslant 3}$ and $(1, -x/t)$.

Proposition 2.3.1 (Hyperboloidal energy estimate). *Given a constant $\kappa_1 > 1$ and some locally integrable functions $L, M \geqslant 0$, the following property holds. Let $(w_i)_{1 \leqslant i \leqslant n_0}$ be the (local-in-time) solution to (2.3.1) defined on some hyperbolic time interval $[s_0, s_1]$ and suppose that the metric and source satisfy the following coercivity conditions*

$$\kappa_1^{-2} \sum_i E_{m,c_i}(s, w_i) \leqslant \sum_i E_{G,c_i}(s, w_i) \leqslant \kappa_1^2 \sum_i E_{m,c_i}(s, w_i), \tag{2.3.5}$$

$$\begin{aligned} &\left| \int_{\mathcal{H}_s} \frac{s}{t} \left(\partial_\alpha G_i^{j\alpha\beta} \partial_t w_i \partial_\beta w_j - \frac{1}{2} \partial_t G_i^{j\alpha\beta} \partial_\alpha w_i \partial_\beta w_j \right) dx \right| \\ &\leqslant M(s) E_{m,c_i}(s, w_i)^{1/2}, \end{aligned} \tag{2.3.6}$$

and the bound

$$\sum_i \|F_i\|_{L^2(\mathcal{H}_s)} \leqslant L(s), \qquad s \in [s_0, s_1]. \tag{2.3.7}$$

Recall also that the symmetry conditions (1.2.2) *are assumed, that is, $G_i^{j\alpha\beta} = G_j^{i\alpha\beta} = G_i^{j\beta\alpha}$, and that F_i, $G_i^{j\alpha\beta}$ are supported in $\mathcal{K}_{[s_0,s_1]}$. The following energy estimate holds (for all $s \in [s_0, s_1]$):*

$$\begin{aligned} &\Big(\sum_i E_{m,c_i}(s, w_i) \Big)^{1/2} \\ &\leqslant \kappa_1^2 \Big(\sum_i E_{m,c_i}(s_0, w_i) \Big)^{1/2} + C\kappa_1^2 \int_{s_0}^s \big(L(\tau) + M(\tau)\big)\, d\tau, \end{aligned} \tag{2.3.8}$$

where the constant $C > 0$ depends on the the structure of the system (2.3.1) *only.*

The following remarks are in order:

- In view of (2.3.2) and (2.3.8), the L^2 norm of $\underline{\partial}_a w_i$ and $(s/t)\partial_\alpha w_i$ are uniformly controlled on each hypersurface $\mathcal{H}_s$. It is expected that these weighted expressions enjoy better decay than $\partial_\alpha w_i$ itself.
- In view of $\sum_a (x^a/r)\big((r/t)\partial_a + (x^a/r)\partial_t\big) w_i = t^{-1} S w_i$, we see that the weighted scaling derivative $t^{-1} S w_i$ is also controlled.
- In Chapter 6, we will see that the energy $E_m(s_0, w_i)$ on the initial hyperboloid is controlled by the $\mathbf{H}^1$ norm of the initial data on the initial slice $t = s_0$.

Proof. In view of the symmetry (1.2.2) and by using $\partial_t w_i$ as a multiplier, we easily derive the energy identity

$$\begin{aligned}
&\sum_i \frac{1}{2}\partial_t\Big(\sum_\alpha(\partial_\alpha w_i)^2 + c_i^2 w_i^2\Big) - \sum_i\sum_a \partial_a(\partial_a w_i \partial_t w_i)\\
&+\sum_i \partial_\alpha(G_i^{j\alpha\beta}\partial_t w_i \partial_\beta w_j) - \sum_i \frac{1}{2}\partial_t(G_i^{j\alpha\beta}\partial_\alpha w_i\partial_\beta w_j)\\
&=\sum_i \partial_t w_i F_i + \sum_i\Big(\partial_\alpha G_i^{j\alpha\beta}\partial_t w_i\partial_\beta w_j - \frac{1}{2}\partial_t G_i^{j\alpha\beta}\partial_\alpha w_i \partial_\beta w_j\Big).
\end{aligned}$$

We integrate this identity over the region $\mathcal{K}_{[s_0,s]}$ and use Stokes' formula. Note that by the property of propagation at finite speed, the solution (w_i) is defined in $\mathcal{K}_{[s_0,+\infty)}$ and vanishing in a neighborhood of the light cone. By our assumption on the support of $G_i^{j\alpha\beta}$, we obtain

$$\begin{aligned}
&\frac{1}{2}\sum_i\big(E_{G,c_i}(s,w_i) - E_{G,c_i}(s_0,w_i)\big)\\
&=\sum_i\int_{\mathcal{K}_{[s_0,s]}}\Big(\partial_t w_i F_i + \partial_\alpha G_i^{j\alpha\beta}\partial_t w_i\partial_\beta w_j - \frac{1}{2}\partial_t G_i^{j\alpha\beta}\partial_\alpha w_i\partial_\beta w_j\Big)\,dtdx\\
&=\sum_i\int_{s_0}^s\Big(\int_{\mathcal{H}_\tau}(\tau/t)\partial_t w_i F_i\,dx\Big)d\tau\\
&\quad+\sum_i\int_{s_0}^s\Big(\int_{\mathcal{H}_\tau}(\tau/t)\Big(\partial_\alpha G_i^{j\alpha\beta}\partial_t w_i\partial_\beta w_j - \frac{1}{2}\partial_t G_i^{j\alpha\beta}\partial_\alpha w_i\partial_\beta w_j\Big)\,dx\Big)d\tau,
\end{aligned}$$

which leads us to

$$\begin{aligned}
\frac{d}{ds}\sum_i E_{G,c_i}(s,w_i) =&2\sum_i\int_{\mathcal{H}_s}\Big((s/t)\partial_\alpha G_i^{j\alpha\beta}\partial_t w_i\partial_\beta w_j\\
&- (s/2t)\partial_t G_i^{j\alpha\beta}\partial_\alpha w_i\partial_\beta w_j + (s/t)\partial_t w_i F_i\Big)\,dx.
\end{aligned}$$

So, with the assumptions (2.3.6) and (2.3.7), we get

$$\begin{aligned}
&\Big(\sum_i E_{G,c_i}(s,w_i)\Big)^{1/2}\frac{d}{ds}\Big(\sum_i E_{G,c_i}(s,w_i)\Big)^{1/2}\\
&\leqslant \sum_i\Big(M(s) + \|F_i\|_{L^2(\mathcal{H}_s)}\Big)E_{m,c_i}(s,w_i)^{1/2}\\
&\leqslant CM(s)\Big(\sum_i E_{m,c_i}(s,w_i)\Big)^{1/2} + C\Big(\sum_i\|F_i\|^2_{L^2(\mathcal{H}_s)}\Big)^{1/2}\Big(\sum_i E_{m,c_i}(s,w_i)\Big)^{1/2}\\
&= C\Big(M(s) + L(s)\Big)\Big(\sum_i E_{m,c_i}(s,w_i)\Big)^{1/2}\\
&\leqslant C\kappa_1\big(M(s)+L(s)\big)\Big(\sum_i E_{G,c_i}(s,w_i)\Big)^{1/2},
\end{aligned}$$

which yields

$$\frac{d}{ds}\Big(\sum_i E_{G,c_i}(s,w_i)\Big)^{1/2} \leqslant C\kappa_1\big(L(s)+M(s)\big).$$

We integrate this inequality over the interval $[s_0, s]$ and obtain

$$\begin{aligned}&\Big(\sum_{i=1}^{n_0} E_{G,c_i}(s,w_i)\Big)^{1/2}\\ &\leqslant \Big(\sum_{i=1}^{n_0} E_{G,c_i}(s_0,w_i)\Big)^{1/2} + C\kappa_1\int_{s_0}^{s} L(\tau)d\tau + C\kappa_1\int_{s_0}^{s} M(\tau)d\tau.\end{aligned}$$

By using the condition (2.3.5), this completes the proof. □

2.4 The bootstrap strategy

We are now in a position to outline our method of proof of Theorem 1.2.1. From now on, we assume that the assumptions therein are satisfied. It will be necessary to distinguish between three levels of regularity and, in order to describe this *scale of regularity*, we will use the following convention on the indices in use:

$$\begin{aligned}&I^\sharp\text{: multi-index of order} \leqslant 5,\\ &I^\dagger\text{: multi-index of order} \leqslant 4,\\ &I\text{: multi-index of order} \leqslant 3,\end{aligned} \tag{2.4.1}$$

which we call **admissible indices.** Throughout, $C, C_0, C_1, C^*, \ldots$ are constants depending only on the structure of the system (1.2.1), such as j_0, k_0, B, c_i.

We will use certain norms on the hyperboloids and, so, if u is a function supported in $\mathcal{K}_{[s_0,+\infty)}$, we set (for $s \geqslant s_0$):

$$\|u\|_{L^p(\mathcal{H}_s)} := \Big(\int_{\mathcal{H}_s} |u(t,x)|^p dx\Big)^{1/p} = \Big(\int_{\mathbb{R}^3} \big|u\big(\sqrt{s^2+|x|^2},x\big)\big|^p dx\Big)^{1/p}. \tag{2.4.2}$$

The proof of Theorem 1.2.1 relies on Propositions 2.4.1–2.4.3, stated now. The first proposition below concerns the construction of initial data on the initial hyperboloid $\mathcal{H}_{B+1}$, and will be established in Chapter 11.

Proposition 2.4.1 (Initialization of the argument). *For any sufficiently large constant $C_0 > 0$, there exists a positive constant $\epsilon_0' \in (0,1)$ depending only on B and C_0 such that, for every initial data satisfying*

$$\sum_i \big(\|w_{i0}\|_{H^6(\mathbb{R}^3)} + \|w_{i1}\|_{H^5(\mathbb{R}^3)}\big) \leqslant \epsilon \leqslant \epsilon_0', \tag{2.4.3}$$

the local-in-time solution to (1.2.1) *associated with this initial data extends to the region limited by the constant time hypersurface* $t = B+1$ *and the hyperboloid* $\mathcal{H}_{B+1}$*. Furthermore, it satisfies the uniform bound*

$$\sum_j E_m(s_0, Z^{I^\sharp} w_j)^{1/2} \leqslant C_0\epsilon. \tag{2.4.4}$$

for all admissible vector fields Z *and all admissible indices* $I^\sharp$.

Given some constants $C_1, \epsilon > 0$ and $\delta \in (0, 1/6)$ and a hyperbolic time interval $[s_0, s_1]$, we call **hierarchy of energy bounds** with parameters (C_1, ϵ, δ) the following five inequalities (for all $s \in [s_0, s_1]$ and all admissible fields and admissible indices):

$$E_m(s, Z^{I^\sharp} u_{\hat\imath})^{1/2} \leqslant C_1\epsilon s^\delta \quad \text{for } 1 \leqslant \hat\imath \leqslant j_0, \tag{2.4.5a}$$

$$E_{m,\sigma}(s, Z^{I^\sharp} v_{\check\jmath})^{1/2} \leqslant C_1\epsilon s^\delta \quad \text{for } j_0+1 \leqslant \check\jmath \leqslant n_0, \tag{2.4.5b}$$

$$E_m(s, Z^{I^\dagger} u_{\hat\imath})^{1/2} \leqslant C_1\epsilon s^{\delta/2} \quad \text{for } 1 \leqslant \hat\imath \leqslant j_0, \tag{2.4.5c}$$

$$E_{m,\sigma}(s, Z^{I^\dagger} v_{\check\jmath})^{1/2} \leqslant C_1\epsilon s^{\delta/2} \quad \text{for } j_0+1 \leqslant \check\jmath \leqslant n_0, \tag{2.4.5d}$$

$$E_m(s, Z^{I} u_{\hat\imath})^{1/2} \leqslant C_1\epsilon \quad \text{for } 1 \leqslant \hat\imath \leqslant j_0. \tag{2.4.5e}$$

Observe that (2.4.5e) concerns the wave components only and that the upper bound is independent of time.

At this juncture, since $c_{\check\imath} \geqslant \sigma > 0$ for all $j_0+1 \leqslant \check\imath \leqslant n_0$, we have

$$E_{m,\sigma}(s, Z^{I^\sharp} v_{\check\imath}) \leqslant E_{m,c_{\check\imath}}(s, Z^{I^\sharp} v_{\check\imath}) \leqslant (c_{\check\imath}/\sigma)^2 E_{m,\sigma}(s, Z^{I^\sharp} v_{\check\imath}), \tag{2.4.6}$$

so that the two energy expressions are equivalent.

Given some constants $\kappa_1 > 1$, $C^*, C_1, \epsilon > 0$ and $\delta \in (0, 1/6)$ and a hyperbolic time interval $[s_0, s_1]$, we call **hierarchy of metric-source bounds** with parameters $(\kappa_1, C^*, C_1, \epsilon, \delta)$. the following three sets of estimates:

- For all $|I^\sharp| \leqslant 5$,

$$\kappa_1^{-2} \sum_i E_{G,c_i}(s, Z^{I^\sharp} w_i) \leqslant \sum_i E_{m,c_i}(s, Z^{I^\sharp} w_i) \leqslant \kappa_1^2 \sum_i E_{G,c_i}(s, Z^{I^\sharp} w_i), \tag{2.4.7a}$$

$$\begin{aligned}
&\left| \int_{\mathcal{H}_s} \frac{s}{t} \left(\partial_\alpha G_i^{j\alpha\beta} \partial_t Z^{I^\sharp} w_i \partial_\beta Z^{I^\sharp} w_j - \frac{1}{2} \partial_t G_i^{j\alpha\beta} \partial_\alpha Z^{I^\sharp} w_i \partial_\beta Z^{I^\sharp} w_j \right) dx \right| \\
&\leqslant C^*(C_1\epsilon)^2 s^{-1+\delta} E_{m,c_i}(s, Z^{I^\sharp} w_i)^{1/2} \\
&=: M(I^\sharp, s) E_{m,c_i}(s, Z^{I^\sharp} w_i)^{1/2},
\end{aligned} \tag{2.4.7b}$$

and*

$$\begin{aligned}&\sum_{i=1}^{n_0} \big\|[G_i^{j\alpha\beta}\partial_\alpha\partial_\beta, Z^{I^\sharp}]w_j\big\|_{L^2(\mathcal{H}_s)} + \sum_{i=1}^{n_0} \big\|Z^{I^\sharp} F_i\big\|_{L^2(\mathcal{H}_s)} \\ &\leqslant C^*(C_1\epsilon)^2 s^{-1+\delta} =: L(I^\sharp, s).\end{aligned} \tag{2.4.7c}$$

- For all $|I^\dagger| \leqslant 4$,

$$\begin{aligned}&\left|\int_{\mathcal{H}_s} \frac{s}{t}\Big(\partial_\alpha G_i^{j\alpha\beta}\partial_t Z^{I^\dagger} w_i \partial_\beta Z^{I^\dagger} w_{\hat{\jmath}} - \frac{1}{2}\partial_t G_i^{j\alpha\beta}\partial_\alpha Z^{I^\dagger} w_i \partial_\beta Z^{I^\dagger} w_j\Big)dx\right| \\ &\leqslant C^*(C_1\epsilon)^2 s^{-1+\delta/2} E_{m,c_i}(s, Z^{I^\dagger} w_i)^{1/2} \\ &=: M(I^\dagger, s) E_{m,c_i}(s, Z^{I^\dagger} w_i)^{1/2}\end{aligned} \tag{2.4.8a}$$

and

$$\begin{aligned}&\sum_{i=1}^{n_0} \big\|[G_i^{j\alpha\beta}\partial_\alpha\partial_\beta, Z^{I^\dagger}]w_j\big\|_{L^2(\mathcal{H}_s)} + \sum_{i=1}^{n_0} \big\|Z^{I^\dagger} F_i\big\|_{L^2(\mathcal{H}_s)} \\ &\leqslant C^*(C_1\epsilon)^2 s^{-1+\delta/2} =: L(I^\dagger, s).\end{aligned} \tag{2.4.8b}$$

- For all $|I| \leqslant 3$,

$$\begin{aligned}&\left|\int_{\mathcal{H}_s} \frac{s}{t}\Big(\partial_\alpha G_{\hat{\imath}}^{j\alpha\beta}\partial_t Z^I u_{\hat{\imath}} \partial_\beta Z^I u_{\hat{\jmath}} - \frac{1}{2}\partial_t G_{\hat{\imath}}^{j\alpha\beta}\partial_\alpha Z^I u_{\hat{\imath}} \partial_\beta Z^I u_{\hat{\jmath}}\Big)dx\right| \\ &\leqslant C^*(C_1\epsilon)^2 s^{-3/2+2\delta} E_m(s, Z^I u_{\hat{\imath}})^{1/2} \\ &=: M(I, s) E_m(s, Z^I u_{\hat{\imath}})^{1/2}\end{aligned} \tag{2.4.9a}$$

and

$$\begin{aligned}&\sum_{i=1}^{n_0} \big\|[G_i^{j\alpha\beta}\partial_\alpha\partial_\beta, Z^I]w_j\big\|_{L^2(\mathcal{H}_s)} \\ &\quad + \sum_{i=1}^{n_0} \big\|Z^I F_i\big\|_{L^2(\mathcal{H}_s)} + \big\|Z^I\big(G_{\hat{\imath}}^{\check{\jmath}\alpha\beta}\partial_\alpha\partial_\beta v_{\check{\jmath}}\big)\big\|_{L^2(\mathcal{H}_s)} \\ &\leqslant C^*(C_1\epsilon)^2 s^{-3/2+2\delta} =: L(I, s).\end{aligned} \tag{2.4.9b}$$

Observe that (2.4.9a) concerns the wave components, only.

Proposition 2.4.2. *Fix some $\delta \in (0, 1/6)$. There exists a positive constant ϵ_0'' such that for all $\epsilon, C_1 > 0$ with $C_1\epsilon \leqslant 1$ and $C_1\epsilon \leqslant \epsilon_0''$, there exists a constant $\kappa_1 > 1$ and a constant $C^* > 0$ (both determined by the structure of the system* (1.2.1)*) such that the hierarchy of energy bounds* (2.4.5) *with parameters C_1, ϵ, δ implies the hierarchy of metric–source bounds* (2.4.7a)—(2.4.9b) *with parameters $\kappa_1, C^*, C_1, \epsilon, \delta$.*

*with the notation (2.4.2)

The proof of this proposition will occupy a major part of this monograph, especially Chapters 7 to 10. Now we admit this proposition and give the proof of the main result, which is going to be essentially based on the following observation.

Proposition 2.4.3 (Enhancing the hierarchy of energy bounds). *Let $\delta \in (0, 1/6)$ and let $C_0 > 0$ be a constant and $C_1 > C_0$ be a sufficiently large constant. Then, there exists $\epsilon_1 > 0$ such that the following enhancing property holds. For any solution (w_i) to (1.2.1) defined in $\mathcal{K}_{[s_0,s_1]}$ and provided, for $\epsilon \in [0, \epsilon_1]$,*

(1) $\sum_{i=1}^{n_0} E_{m,c_i}(s_0, Z^{I^\sharp} w_i)^{1/2} \leqslant C_0\epsilon$ for all admissible Z and $I^\sharp$,
(2) the hierarchy (2.4.5) holds with parameters (C_1, ϵ, δ) in the time interval $[s_0, s_1]$,

then the following **improved energy estimates** *also hold*

$$E_m(s, Z^{I^\sharp} u_{\hat{\imath}})^{1/2} \leqslant \frac{1}{2} C_1 \epsilon s^\delta \quad \text{for } 1 \leqslant \hat{\imath} \leqslant j_0, \tag{2.4.10a}$$

$$E_{m,\sigma}(s, Z^{I^\sharp} v_{\check{\jmath}})^{1/2} \leqslant \frac{1}{2} C_1 \epsilon s^\delta \quad \text{for } j_0 + 1 \leqslant \check{\jmath} \leqslant n_0, \tag{2.4.10b}$$

$$E_m(s, Z^{I^\dagger} u_{\hat{\imath}})^{1/2} \leqslant \frac{1}{2} C_1 \epsilon s^{\delta/2} \quad \text{for } 1 \leqslant \hat{\imath} \leqslant j_0, \tag{2.4.10c}$$

$$E_{m,\sigma}(s, Z^{I^\dagger} v_{\check{\jmath}})^{1/2} \leqslant \frac{1}{2} C_1 \epsilon s^{\delta/2} \quad \text{for } j_0 + 1 \leqslant \check{\jmath} \leqslant n_0, \tag{2.4.10d}$$

$$E_m(s, Z^{I} u_{\hat{\imath}})^{1/2} \leqslant \frac{1}{2} C_1 \epsilon \quad \text{for } 1 \leqslant \hat{\imath} \leqslant j_0. \tag{2.4.10e}$$

Proof. We will use here the conclusion of Proposition 2.4.2. We assume that ϵ_1 is sufficiently small with $\epsilon_1 \leqslant C_1^{-1} \min\{1, \epsilon_0''\}$ such that, for all $\epsilon \leqslant \epsilon_1$, the conclusion in Proposition 2.4.2 holds.

Step I. High-order energy estimates. To the equations (1.2.1), we apply the operator $Z^{I^\sharp}$ (with $|I^\sharp| \leqslant 5$ throughout) and obtain

$$\Box(Z^{I^\sharp} w_i) + G_i^{j\alpha\beta} \partial_\alpha \partial_\beta (Z^{I^\sharp} w_j) + c_i^2 Z^{I^\sharp} w_i = [G_i^{j\alpha\beta} \partial_\alpha \partial_\beta, Z^{I^\sharp}] w_j + Z^{I^\sharp} F_i.$$

In view of Proposition 2.3.1 and thanks to the conditions (2.4.7a)–(2.4.7c) implied by Proposition 2.4.2, we find

$$\begin{aligned}
&\Big(\sum_i E_{m,c_i}(s, Z^{I^\sharp} w_i)\Big)^{1/2} \\
&\leqslant \kappa_1^2 \Big(\sum_i E_{m,c_i}(s_0, Z^{I^\sharp} w_i)\Big)^{1/2} + C\kappa_1^2 \int_{s_0}^s \big(L(I^\sharp, \tau) + M(I^\sharp, \tau)\big) d\tau.
\end{aligned}$$

In view of Proposition 2.4.2 combined with the condition (1) in the proposition, we then have

$$\Big(\sum_i E_{m,c_i}(s, Z^{I^\sharp} w_i)\Big)^{1/2} \leqslant \kappa_1^2 C_0 \epsilon + C\kappa_1^2 C^* (C_1\epsilon)^2 \int_{s_0}^s \tau^{-1+\delta} d\tau.$$

Choosing $\epsilon_1 \leqslant \frac{\delta(C_1 - 2\kappa_1^2 C_0)}{2CC^*\kappa_1^2 C_1^2}$, we obtain

$$\Big(\sum_i E_{m,c_i}(s, Z^{I^\sharp} w_i)\Big)^{1/2} \leqslant \frac{1}{2} C_1 \epsilon s^\delta,$$

which leads to the enhanced energy bound

$$E_{m,\sigma}(s, Z^{I^\sharp} w_i)^{1/2} \leqslant E_{m,c_i}(s, Z^{I^\sharp} w_i)^{1/2} \leqslant \frac{1}{2} C_1 \epsilon s^\delta. \tag{2.4.11}$$

This establishes (2.4.10b)-(2.4.10a).

Step II. Intermediate energy estimates. We will next rely on the metric–source bounds (2.4.8a)-(2.4.8b) (with $|I^\dagger| \leqslant 4$ throughout) given by the conclusion of Proposition 2.4.2. To the equation (1.2.1), we apply the operator $Z^{I^\dagger}$ and obtain

$$\Box(Z^{I^\dagger} w_i) + G_i^{j\alpha\beta} \partial_\alpha \partial_\beta (Z^{I^\dagger} w_j) + c_i^2 Z^{I^\dagger} w_i = [G_i^{j\alpha\beta} \partial_\alpha \partial_\beta, Z^{I^\dagger}] w_j + Z^{I^\dagger} F_i.$$

From Proposition 2.3.1 and thanks to the conditions (2.4.7a) and (2.4.8a)-(2.4.8b), we find

$$\begin{aligned} &\Big(\sum_i E_{m,c_i}(s, Z^{I^\dagger} w_i)\Big)^{1/2} \\ &\leqslant \kappa_1^2 \Big(\sum_i E_{m,c_i}(s_0, Z^{I^\dagger} w_i)\Big)^{1/2} + C\kappa_1^2 \int_{s_0}^s \big(L(I^\dagger, \tau) + M(I^\dagger, \tau)\big) d\tau. \end{aligned}$$

We thus have

$$\Big(\sum_i E_{m,c_i}(s, Z^{I^\dagger} w_i)\Big)^{1/2} \leqslant \kappa_1^2 C_0 \epsilon + C\kappa_1^2 C^* (C_1\epsilon)^2 \int_{s_0}^s \tau^{-1+\delta/2} d\tau.$$

Provided ϵ is sufficiently small so that $\epsilon_1 \leqslant \frac{\delta(C_1 - 2\kappa_1^2 C_0)}{4CC^*\kappa_1^2 C_1^2}$, we find

$$\Big(\sum_i E_{m,c_i}(s, Z^{I^\dagger} w_i)\Big)^{1/2} \leqslant \frac{1}{2} C_1 \epsilon s^{\delta/2},$$

which leads to the enhanced energy bound

$$E_{m,\sigma}(s, Z^{I^\dagger} w_i)^{1/2} \leqslant E_{m,c_i}(s, Z^{I^\dagger} w_i)^{1/2} \leqslant \frac{1}{2} C_1 \epsilon s^{\delta/2}. \tag{2.4.12}$$

This establishes (2.4.10d)-(2.4.10c).

Step III. Low-order energy estimates. We will now rely on the metric–source bounds (2.4.9a)-(2.4.9b) (with $|I| \leqslant 3$ throughout). We apply Z^I to the wave equations in (1.2.1) and obtain

$$\Box(Z^I u_{\widehat{\imath}}) + G_{\widehat{\imath}}^{\widehat{\jmath}\alpha\beta}\partial_{\alpha\beta}(Z^I u_{\widehat{\jmath}}) = [G_{\widehat{\imath}}^{\widehat{\jmath}\alpha\beta}\partial_\alpha\partial_\beta, Z^I]u_{\widehat{\jmath}} - Z^I\big(G_{\widehat{\imath}}^{\check{\jmath}\alpha\beta}\partial_\alpha\partial_\beta v_{\check{\jmath}}\big) + Z^I F_{\widehat{\imath}}.$$

In view of Proposition 2.3.1 and thanks to the conditions (2.4.7a) and (2.4.9a)-(2.4.9b), we obtain

$$\begin{aligned}&\Big(\sum_{\widehat{\imath}} E_m(s, Z^I u_{\widehat{\imath}})\Big)^{1/2} \\ &\leqslant \kappa_1^2\Big(\sum_{\widehat{\imath}} E_m(s_0, Z^I u_{\widehat{\imath}})\Big)^{1/2} + C\kappa_1^2\int_{s_0}^s \big(L(I,\tau) + M(I,\tau)\big)\, d\tau\end{aligned}$$

and, therefore,

$$\Big(\sum_{\widehat{\imath}} E_m(s, Z^I u_{\widehat{\imath}})\Big)^{1/2} \leqslant \kappa_1^2 C_0\epsilon + C\kappa_1^2 C^*(C_1\epsilon)^2\int_{s_0}^s \tau^{-3/2+2\delta}d\tau.$$

By recalling that $\delta < 1/6$, it follows that

$$\Big(\sum_{\widehat{\imath}} E_m(s, Z^I u_{\widehat{\imath}})\Big)^{1/2} \leqslant \kappa_1^2 C_0\epsilon + \frac{CC^*\kappa_1^2 C_1^2\epsilon^2(B+1)^{2\delta-1/2}}{(1/2)-2\delta}.$$

Provided $\epsilon_1 \leqslant \frac{(1-4\delta)(C_1-2\kappa_1^2C_0)}{4CC^*\kappa_1^2C_1^2(B+1)^{2\delta-1/2}}$, we obtain the enhanced energy bound for the wave components

$$E_m(s, Z^I u_{\widehat{\imath}})^{1/2} \leqslant \frac{1}{2}C_1\epsilon$$

and this establishes (2.4.10e).

Step IV. In conclusion, by assuming that

$$\epsilon_1 \leqslant \min\Big(\frac{\delta(C_1 - 2\kappa_1^2 C_0)}{4CC^*\kappa_1^2C_1^2}, \frac{(1-4\delta)(C_1-2\kappa_1^2C_0)}{4CC^*\kappa_1^2C_1^2(B+1)^{2\delta-1/2}}, \epsilon_0'', C_1^{-1}\Big),$$

all conditions (2.4.10) are satisfied and the proof is completed. □

Observe that (2.4.7a), (2.4.7b), (2.4.8a), and (2.4.9a) concern only the "metric" $G_i^{j\alpha\beta}$ and depend mainly on L^∞ bounds on the solution and its derivatives: these bounds will be established in Chapter 9. On the other hand, the inequalities (2.4.7c), (2.4.8b), and (2.4.9b) are L^2 type estimates and will be derived later in Chapter 10.

We complete this chapter with a proof of our main result —by assuming that Propositions 2.4.1 and 2.4.2 are established.

Proof of Theorem 1.2.1. Let (w_i) be the unique local-in-time solution to (1.2.1) associated with the initial data (w_{i0}, w_{i1}). Given our assumption on the support of (w_{i0}, w_{i1}) and according to the property of propagation at finite speed, this solution (w_i) must be supported in the region $K_{[s_0,+\infty)}$.

As guaranteed by Proposition 2.4.1, there exist constants C_0, ϵ_0' such that, provided $\|w_{i0}\|_{H^6(\mathbb{R}^3)} + \|w_{i1}\|_{H^5(\mathbb{R}^3)} \leqslant \epsilon \leqslant \epsilon_0'$, we have the energy bound $E_G(s_0, Z^{I^\sharp} w_j)^{1/2} \leqslant C_0\epsilon$.

Let $[s_0, s^*]$ be the largest time interval (containing s_0) on which (2.4.5) holds with some parameters $C_1, \epsilon > 0$ and $\delta \in (0, 1/6)$ fixed and for a sufficiently large constant $C_1 > C_0$. By continuity, we have $s^* > s_0$.

Proceeding by contradiction, let us assume that $s^* < +\infty$, so that at the time $s = s^*$, one (at least) of the inequalities (2.4.5) must be an equality. That is, at least one of the following conditions holds:

$$
\begin{aligned}
&E_{m,\sigma}(s^*, Z^{I^\sharp} v_{\check{\jmath}})^{1/2} = C_1\epsilon s^{*\delta} \quad \text{for } j_0+1 \leqslant \check{\jmath} \leqslant n_0,\\
&E_m(s^*, Z^{I^\sharp} u_{\hat{\imath}})^{1/2} = C_1\epsilon s^{*\delta} \quad \text{for } 1 \leqslant \hat{\imath} \leqslant j_0,\\
&E_{m,\sigma}(s^*, Z^{I^\dagger} v_{\check{\jmath}})^{1/2} = C_1\epsilon s^{*\delta/2} \quad \text{for } j_0+1 \leqslant \check{\jmath} \leqslant n_0,\\
&E_m(s^*, Z^{I^\dagger} u_{\hat{\imath}})^{1/2} = C_1\epsilon s^{*\delta/2} \quad \text{for } 1 \leqslant \hat{\imath} \leqslant j_0,\\
&E_m(s^*, Z^{I} u_{\hat{\imath}})^{1/2} = C_1\epsilon, \quad \text{for } 1 \leqslant \hat{\imath} \leqslant j_0.
\end{aligned}
\tag{2.4.13}
$$

Yet, according to Proposition 2.4.3 there exists $\epsilon_1 > 0$ such that, for $\epsilon \leqslant \epsilon_1$, the following enhanced energy bounds also hold:

$$
\begin{aligned}
&E_{m,\sigma}(s^*, Z^{I^\sharp} v_{\check{\jmath}})^{1/2} \leqslant \frac{1}{2}C_1\epsilon s^{*\delta} \quad \text{for } j_0+1 \leqslant \check{\jmath} \leqslant n_0,\\
&E_m(s^*, Z^{I^\sharp} u_{\hat{\imath}})^{1/2} \leqslant \frac{1}{2}C_1\epsilon s^{*\delta} \quad \text{for } 1 \leqslant \hat{\imath} \leqslant j_0,\\
&E_{m,\sigma}(s^*, Z^{I^\dagger} v_{\check{\jmath}})^{1/2} \leqslant \frac{1}{2}C_1\epsilon s^{*\delta/2} \quad \text{for } j_0+1 \leqslant \check{\jmath} \leqslant n_0,\\
&E_m(s^*, Z^{I^\dagger} u_{\hat{\imath}})^{1/2} \leqslant \frac{1}{2}C_1\epsilon s^{*\delta/2} \quad \text{for } 1 \leqslant \hat{\imath} \leqslant j_0,\\
&E_m(s^*, Z^{I} u_{\hat{\imath}})^{1/2} \leqslant \frac{1}{2}C_1\epsilon \quad \text{for } 1 \leqslant \hat{\imath} \leqslant j_0.
\end{aligned}
$$

Clearly, this is impossible, unless $s^* = +\infty$.

Hence, we have proven that the inequalities (2.4.5) hold for all $s \in [s_0, +\infty)$ and, by the local existence criteria in Theorem 11.2.1 (combined with Sobolev's inequality), this local solution extends for all times. By taking $\epsilon_0 := \min(\epsilon_0', \epsilon_1)$, this completes the proof of Theorem 1.2.1. □

2.5 Energy on the hypersurfaces of constant time

To end this chapter, we prove that the wave components of the global-in-time solutions which we have constructed for (1.2.1) enjoy a uniform energy estimate also on the standard hypersurfaces of constant time t.

Proposition 2.5.1. *Consider the wave components u_i of a global-in-time solution w_i satisfying (2.4.5) and (2.4.9) (with ϵ_0 sufficiently small and $C_1\epsilon < 1$). The following estimate also holds for all time $t \geqslant B+1$:*

$$\begin{aligned}&\|\partial_t Z^I u(t,\cdot)\|_{L^2(\mathbb{R}^3)} + \sum_a \|\partial_a Z^I u(t,\cdot)\|_{L^2(\mathbb{R}^3)}\\ &\leqslant C_3(B,\delta,\epsilon_0)C_1\epsilon,\end{aligned} \tag{2.5.1}$$

where C_3 depends upon δ, ϵ_0 and the structure of the system only.

The proof of this result relies on a modified version of the fundamental energy estimate, as now stated.

Lemma 2.5.1. *Let (u_i) be the solution to the Cauchy problem*

$$\begin{aligned}&\Box u_i + G_i^{j\alpha\beta}\partial_{\alpha\beta}u_i = F_i,\\ &u_i|_{\mathcal{H}_{s0}} = u_{i0}, \quad \partial_t u_i|_{\mathcal{H}_{s0}} = u_{i0},\end{aligned} \tag{2.5.2}$$

where F_i and $G_i^{j\alpha\beta}$ are defined in $\mathcal{K}_{[s_0,+\infty)}$, while $u_i|_{\mathcal{H}_{s_0}}, \partial_t u_i|_{\mathcal{H}_{s_0}}$ are supported in $\mathcal{H}_{s_0}\cap\mathcal{K}$. There exists $\varepsilon_0 > 0$ such that, provided

$$\max_{i,j,\alpha,\beta}|G_i^{j\alpha\beta}| \leqslant \varepsilon_0,$$

there exists a positive constant $C = C(\epsilon_0)$ such that

$$\begin{aligned}&\Big(\|\partial_t u(t_0,\cdot)\|_{L^2(\mathbb{R}^3)} + \sum_a \|\partial_a u(t_0,\cdot)\|_{L^2(\mathbb{R}^3)}\Big)^2\\ &\leqslant C\sum_i E_G(t,u_i) + C\sum_i \int_{t_0}^{(t_0^2+1)/2}\int_{\mathcal{H}_s}(s/t)\big|\partial_t u_i F_i\big|\,dxds\\ &\quad + C\int_{t_0}^{(t_0^2+1)/2}\int_{\mathcal{H}_s}(s/t)\big|\partial_\alpha G_i^{j\alpha\beta}\partial_t u_i\partial_\beta u_j - \frac{1}{2}\partial_t G_i^{j\alpha\beta}\partial_\alpha u_i\partial_\beta u_j\big|\,dxds.\end{aligned} \tag{2.5.3}$$

Proof. The proof relies on an energy estimate within the region $\mathcal{K}_{t_0} := \{(t,x)\,|\,|x| \leqslant t-1, t_0 \leqslant t \leqslant \sqrt{|x|^2+t_0^2}\}$, which we display in Fig. 2.3. We

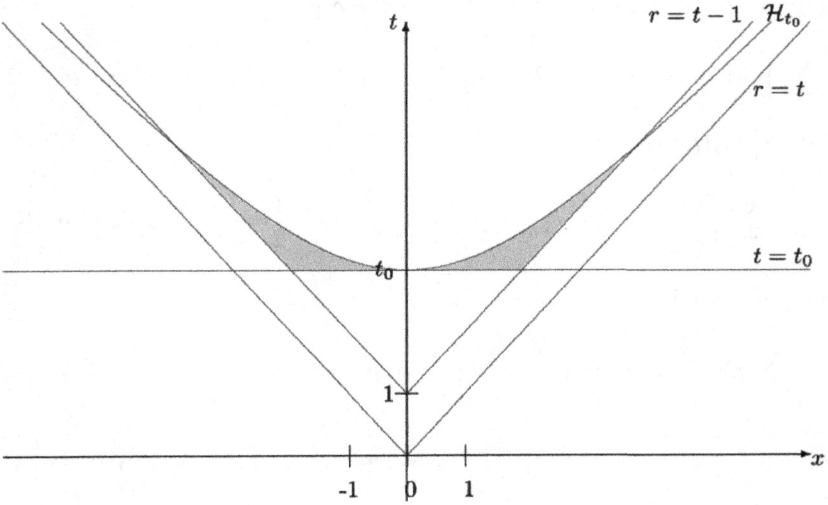

Fig. 2.3 The region $\mathcal{K}_{t_0}$

begin from the general identity

$$\begin{aligned}
&\sum_i \Big(\frac{1}{2}\partial_t \sum_\alpha \big(\partial_\alpha u_i\big)^2 + \sum_a \partial_a\big(\partial_a u_i \partial_t u_i\big) \\
&\qquad + \partial_\alpha\big(G_i^{j\alpha\beta}\partial_t u_i \partial_\beta u_j\big) - \frac{1}{2}\partial_t\big(G_i^{j\alpha\beta}\partial_\alpha u_i \partial_\beta u_j\big)\Big) \\
&= \sum_i \partial_t u_i F_i + \sum_i \Big(\partial_\alpha G_i^{j\alpha\beta}\partial_t u_i \partial_\beta u_j - \frac{1}{2}\partial_t G_i^{j\alpha\beta}\partial_\alpha u_i \partial_\beta u_j\Big)
\end{aligned}$$

and we integrate it within $\mathcal{K}_{t_0}$ with respect to the volume form $dtdx$.

Applying Stokes' formula, we find

$$\begin{aligned}
&\frac{1}{2}\sum_i E_G(t, u_i) \\
&- \frac{1}{2}\sum_i \int_{\mathbb{R}^3} \Big(\sum_\alpha |\partial_\alpha u_i|^2 + 2G_i^{j0\beta}\partial_t u_i \partial_\beta u_j - G_i^{j\alpha\beta}\partial_\alpha u_i \partial_\beta u_j\Big)(t, \cdot)dx \\
&= \int_{\mathcal{K}_{t_0}} \sum_i \Big(\partial_t u_i F_i + \partial_\alpha G_i^{j\alpha\beta}\partial_t u_i \partial_\beta u_j - \frac{1}{2}\partial_t G_i^{j\alpha\beta}\partial_\alpha u_i \partial_\beta u_j\Big)\, dxdt \\
&= \int_{\mathcal{K}_{t_0}} (s/t)\sum_i \Big(\partial_t u_i F_i + \partial_\alpha G_i^{j\alpha\beta}\partial_t u_i \partial_\beta u_j - \frac{1}{2}\partial_t G_i^{j\alpha\beta}\partial_\alpha u_i \partial_\beta u_j\Big)\, dxds,
\end{aligned}$$

where we recall that $s = \sqrt{t^2 - |x|^2}$. This implies

$$\begin{aligned}
&\sum_i \int_{\mathbb{R}^3} \Big(\sum_\alpha |\partial_\alpha u_i|^2 + 2G_i^{j0\beta} \partial_t u_i \partial_\beta u_j - G_i^{j\alpha\beta} \partial_\alpha u_i \partial_\beta u_j \Big)(t,\cdot) dx \\
&\leqslant \sum_i E_G(t, u_i) \\
&\quad + \sum_i \int_{\mathcal{K}_{t_0}} (s/t) \big| \partial_t u_i F_i + \partial_\alpha G_i^{j\alpha\beta} \partial_t u_i \partial_\beta u_j - \frac{1}{2} \partial_t G_i^{j\alpha\beta} \partial_\alpha u_i \partial_\beta u_j \big| \, dx ds,
\end{aligned}$$

which is bounded by

$$\begin{aligned}
&\sum_i E_G(t, u_i) + \sum_i \int_{t_0}^{(t_0^2+1)/2} \int_{\mathcal{H}_s} (s/t) \big| \partial_t u_i F_i \big| \, dx ds \\
&+ \int_{t_0}^{(t_0^2+1)/2} \int_{\mathcal{H}_s} (s/t) \big| \partial_\alpha G_i^{j\alpha\beta} \partial_t u_i \partial_\beta u_j - \frac{1}{2} \partial_t G_i^{j\alpha\beta} \partial_\alpha u_i \partial_\beta u_j \big| \, dx ds.
\end{aligned}$$

We then observe that

$$\begin{aligned}
&\sum_i \int_{\mathbb{R}^3} \big| 2G_i^{j0\beta} \partial_t u_i \partial_\beta u_j - G_i^{j\alpha\beta} \partial_\alpha u_i \partial_\beta u_j \big| \, dx \\
&\leqslant C \max_{i,j,\alpha,\beta} |G_i^{j\alpha\beta}| \sum_{i,\alpha} \|\partial_\alpha u_i(t,\cdot)\|_{L^2(\mathbb{R}^3)}
\end{aligned}$$

and thus, provided $\max_{i,j,\alpha,\beta} |G_i^{j\alpha\beta}|$ is sufficiently small,

$$\sum_i \int_{\mathbb{R}^3} \big| 2G_i^{j0\beta} \partial_t u_i \partial_\beta u_j - G_i^{j\alpha\beta} \partial_\alpha u_i \partial_\beta u_j \big| \, dx \leqslant \frac{1}{2} \sum_{i,\alpha} \|\partial_\alpha u_i(t,\cdot)\|_{L^2(\mathbb{R}^3)}.$$

We thus have

$$\begin{aligned}
&\frac{1}{2} \sum_{i,\alpha} \|\partial_\alpha u(t,\cdot)\|_{L^2(\mathbb{R}^3)} \\
&\leqslant \sum_i \int_{\mathbb{R}^3} \Big(\sum_\alpha |\partial_\alpha u_i|^2 + 2G_i^{j0\beta} \partial_t u_i \partial_\beta u_j - G_i^{j\alpha\beta} \partial_\alpha u_i \partial_\beta u_j \Big) dx
\end{aligned}$$

and the desired result is proven. □

Proof of Proposition 2.5.1. We recall the equations satisfied by the wave components:

$$\Box(Z^I u_{\hat{\imath}}) + G_{\hat{\imath}}^{\hat{\jmath}\alpha\beta} \partial_{\alpha\beta}(Z^I u_{\hat{\jmath}}) = [G_{\hat{\imath}}^{\hat{\jmath}\alpha\beta} \partial_\alpha \partial_\beta, Z^I] u_{\hat{\jmath}} - Z^I \big(G_{\hat{\imath}}^{\check{\jmath}\alpha\beta} \partial_\alpha \partial_\beta v_{\check{\jmath}} \big) + Z^I F_{\hat{\imath}}$$

and apply (2.5.3), so that

$$
\begin{aligned}
&\sum_{\widehat{\imath},\alpha} \|\partial_\alpha Z^I u_{\widehat{\imath}}(t,\cdot)\|_{L^2(\mathbb{R}^3)} \\
&\leqslant C\sum_{\widehat{\imath}} E_G(t,Z^I u_{\widehat{\imath}}) + C\sum_{\widehat{\imath}}\int_{t_0}^{(t_0^2+1)/2}\int_{\mathcal{H}_s}(s/t)\big|\partial_t Z^I u_{\widehat{\imath}} Z^I F_{\widehat{\imath}}\big|\,dxds \\
&\quad+\sum_{\widehat{\imath}}\int_{t_0}^{(t_0^2+1)/2}\int_{\mathcal{H}_s}(s/t)\big|\partial_t Z^I u_{\widehat{\imath}}[G_{\widehat{\imath}}^{\widehat{\jmath}\alpha\beta}\partial_\alpha\partial_\beta, Z^I]u_{\widehat{\jmath}}\big|\,dxds \\
&\quad+\sum_{\widehat{\imath}}\int_{t_0}^{(t_0^2+1)/2}\int_{\mathcal{H}_s}(s/t)\big|\partial_t Z^I u_{\widehat{\imath}} Z^I\big(G_{\widehat{\imath}}^{\check{\jmath}\alpha\beta}\partial_\alpha\partial_\beta v_{\check{\jmath}}\big)\big|\,dxds \\
&+C\sum_{\widehat{\imath}}\int_{t_0}^{(t_0^2+1)/2}\int_{\mathcal{H}_s}(s/t)\Big|\partial_\alpha G_{\widehat{\imath}}^{j\alpha\beta}\partial_t Z^I u_{\widehat{\imath}}\partial_\beta Z^I u_j - \frac{1}{2}\partial_t G_{\widehat{\imath}}^{j\alpha\beta}\partial_\alpha Z^I u_{\widehat{\imath}}\partial_\beta Z^I u_j\Big|dxds.
\end{aligned}
$$

By (2.4.5e) (which was established within our bootstrap argument and holds on the time interval $[B+1,+\infty)$), we have

$$
\sum_{\widehat{\imath}} E_G(t,Z^I u_{\widehat{\imath}}) \leqslant (C_1\epsilon)^2.
$$

Also, the second term in the right-hand side is uniformly bounded, as follows:

$$
\begin{aligned}
&\int_{t_0}^{(t_0^2+1)/2}\int_{\mathcal{H}_s}(s/t)\big|\partial_t Z^I u_{\widehat{\imath}} Z^I F_{\widehat{\imath}}\big|\,dxds \\
&\leqslant \int_{t_0}^{(t_0^2+1)/2}\|(s/t)\partial_t Z^I u_{\widehat{\imath}}\|_{L^2(\mathcal{H}_s)}\|Z^I F_{\widehat{\imath}}\|_{L^2(\mathcal{H}_s)}ds \\
&\leqslant C_1\epsilon\int_{t_0}^{(t_0^2+1)/2}\|Z^I F_{\widehat{\imath}}\|_{L^2(\mathcal{H}_s)}ds \\
&\leqslant C_1\epsilon\int_{t_0}^{(t_0^2+1)/2} CC^*(C_1\epsilon)^2 s^{-3/2+2\delta}ds \\
&\leqslant CC^*(C_1\epsilon)^3(1/2-2\delta)^{-1}(B+1)^{-1/2+2\delta}.
\end{aligned}
$$

The third and fourth terms are bounded in the same manner, that is,

$$
\begin{aligned}
&\int_{t_0}^{(t_0^2+1)/2}\int_{\mathcal{H}_s}(s/t)\big|\partial_t Z^I u_{\widehat{\imath}}[G_{\widehat{\imath}}^{\widehat{\jmath}\alpha\beta}\partial_\alpha\partial_\beta, Z^I]u_{\widehat{\jmath}}\big|\,dxds \\
&\leqslant CC^*(C_1\epsilon)^3(1/2-2\delta)^{-1}(B+1)^{-1/2+2\delta}
\end{aligned}
$$

and

$$
\begin{aligned}
&\int_{t_0}^{(t_0^2+1)/2}\int_{\mathcal{H}_s}(s/t)\big|\partial_t Z^I u_{\widehat{\imath}} Z^I\big(G_{\widehat{\imath}}^{\check{\jmath}\alpha\beta}\partial_\alpha\partial_\beta v_{\check{\jmath}}\big)\big|\,dxds \\
&\leqslant CC^*(C_1\epsilon)^3(1/2-2\delta)^{-1}(B+1)^{-1/2+2\delta}.
\end{aligned}
$$

The last term is bounded by applying (2.4.9a):

$$\int_{t_0}^{(t_0^2+1)/2}\int_{\mathcal{H}_s}(s/t)\big|\partial_\alpha G_{\hat{\imath}}^{j\alpha\beta}\partial_t Z^I u_{\hat{\imath}}\partial_\beta Z^I u_j-\frac{1}{2}\partial_t G_{\hat{\imath}}^{j\alpha\beta}\partial_\alpha Z^I u_{\hat{\imath}}\partial_\beta Z^I u_j\big|\,dxds$$
$$\leqslant CC^*(C_1\epsilon)^3(1/2-2\delta)^{-1}(B+1)^{-1/2+2\delta}.$$

□

Chapter 3

Decompositions and estimates for the commutators

3.1 Decompositions of commutators. I

In this chapter, we present technical results concerning the commutators $[X,Y]u := X(Yu) - Y(Xu)$ of certain operators X, Y associated with the set of admissible vector fields Z (cf. Chapter 2) and applied to functions u defined in the cone $\mathcal{K} = \{|x| < t-1\}$. In order to derive uniform bounds, we will rely on homogeneity arguments and on the observation that all the coefficients in the following decomposition are smooth within $\mathcal{K}$.

First of all, all admissible vector fields ∂_α, L_a under consideration are Killing fields for the flat wave operator $\Box$, so that the following commutation relations hold:

$$[\partial_\alpha, \Box] = 0, \qquad [L_a, \Box] = 0. \tag{3.1.1}$$

By introducing the notation

$$[L_a, \partial_\beta]u =: \Theta_{a\beta}^{\gamma}\partial_\gamma u, \tag{3.1.2a}$$

$$[\partial_\alpha, \underline{\partial}_\beta]u =: t^{-1}\underline{\Gamma}_{\alpha\beta}^{\gamma}\partial_\gamma u, \tag{3.1.2b}$$

$$[L_a, \underline{\partial}_\beta]u =: \underline{\Theta}_{a\beta}^{\gamma}\underline{\partial}_\gamma u, \tag{3.1.2c}$$

we find easily that

$$\begin{aligned}
&\Theta_{ab}^{\gamma} = -\delta_{ab}\delta_0^{\gamma}, && \Theta_{a0}^{\gamma} = -\delta_a^{\gamma}, && \\
&\underline{\Gamma}_{ab}^{\gamma} = \delta_{ab}\delta_0^{\gamma}, && \underline{\Gamma}_{0b}^{\gamma} = -\frac{x^b}{t}\delta_0^{\gamma} = \Psi_b^0\delta_0^{\gamma}, && \underline{\Gamma}_{\alpha 0}^{\gamma} = 0,\\
&\underline{\Theta}_{ab}^{\gamma} = -\frac{x^b}{t}\delta_a^{\gamma} = \Psi_b^0\delta_a^{\gamma}, && \underline{\Theta}_{a0}^{\gamma} = -\delta_a^{\gamma} + \frac{x^a}{t}\delta_0^{\gamma} = -\delta_a^{\gamma} + \Phi_0^a\delta_0^{\gamma},
\end{aligned} \tag{3.1.3}$$

where Φ and Ψ were defined in (2.2.2). All of these coefficients are smooth in the cone $\mathcal{K}$ and for each index I, the functions $Z^I\Pi$ are bounded for all $\Pi \in \{\Theta, \underline{\Theta}, \underline{\Gamma}\}$. Furthermore, we can also check that

$$\underline{\Theta}_{ab}^{0} = 0 \quad \text{so that} \quad [L_a, \underline{\partial}_b] = \underline{\Theta}_{ab}^{\beta}\underline{\partial}_\beta = \underline{\Theta}_{ab}^{c}\underline{\partial}_c. \tag{3.1.4}$$

Lemma 3.1.1 (Algebraic decompositions of commutators. I). *There exist* constants $\theta_{\alpha J}^{I\beta}$ *such that, for all sufficiently regular function* u *defined in the cone* $\mathcal{K}$, *the following identities hold for all* α *and* I*:*

$$[Z^I, \partial_\alpha]u = \sum_{|J|<|I|} \theta_{\alpha J}^{I\beta}\partial_\beta Z^J u. \tag{3.1.5}$$

Proof. The proof is done by induction on $|I|$. When $|I| = 1$, (3.1.5) is implied by (3.1.2a). Assume next that (3.1.5) is valid for $|I| \leqslant k$ and let us derive this property for all $|I| = k+1$. Let Z^I be a product with index satisfying $|I| = k+1$, where $Z^I = Z_1 Z^{I'}$ with $|I'| = k$ and Z_1 is one of the fields ∂_γ, L_a. In view of

$$[Z^I, \partial_\alpha]u = [Z_1 Z^{I'}, \partial_\alpha]u = Z_1\big([Z^{I'}, \partial_\alpha]u\big) + [Z_1, \partial_\alpha]Z^{I'}u$$

and

$$Z_1\big([Z^{I'}, \partial_\alpha]u\big) = Z_1\Big(\sum_{|J|\leqslant k-1} \theta_{\alpha J}^{I'\gamma}\partial_\gamma Z^J u\Big) = \sum_{|J|\leqslant k-1} \theta_{\alpha J}^{I'\gamma} Z_1\partial_\gamma Z^J u,$$

we have

$$[Z^I, \partial_\alpha]u = \sum_{|J|\leqslant k-1} \theta_{\alpha J}^{I'\gamma} Z_1\partial_\gamma Z^J u + [Z_1, \partial_\alpha]Z^{I'}u.$$

First of all, if $Z_1 = \partial_\beta$, we have the commutation property $[Z_1, \partial_\alpha]Z^{I'}u = 0$ and

$$\sum_{|J|\leqslant k-1} \theta_{\alpha J}^{I'\gamma} Z_1\partial_\gamma Z^J u = \sum_{|J|\leqslant k-1} \theta_{\alpha J}^{I'\gamma}\partial_\gamma Z_1\, Z^J u,$$

so that (3.1.5) is established in this case.

Second, if $Z_1 = L_a$, we apply (3.1.2a) and write

$$\begin{aligned}\sum_{|J|\leqslant k-1} \theta_{\alpha J}^{I_1\gamma} L_a\partial_\gamma Z^J u &= \sum_{|J|\leqslant k-1} \theta_{\alpha J}^{I'\gamma}\partial_\gamma L_a Z^J u + \sum_{|J|\leqslant k-1} \theta_{\alpha J}^{I'\gamma}[L_a, \partial_\gamma]Z^J u \\ &= \sum_{|J|\leqslant k-1} \theta_{\alpha J}^{I'\gamma}\partial_\gamma L_a Z^J u + \sum_{|J|\leqslant k-1} \theta_{\alpha J}^{I'\beta}\Theta_{a\beta}^{\gamma}\partial_\gamma Z^J u.\end{aligned}$$

So, (3.1.5) holds for $|I| = k+1$, which completes the induction argument. □

3.2 Decompositions of commutators. II and III

Lemma 3.2.1 (Algebraic decompositions of commutators. II). *For all sufficiently regular functions u defined in the cone $\mathcal{K}$, the following identity holds:*

$$[Z^I, \underline{\partial}_\beta]u = \sum_{|J|<|I|} \underline{\theta}_{\beta J}^{I\gamma} \partial_\gamma Z^J u, \tag{3.2.1}$$

where the functions $\underline{\theta}_{\beta J}^{I\gamma}$ are smooth and satisfy in $\mathcal{K}$:

$$\left|\partial^{I_1} Z^{I_2} \underline{\theta}_{\beta J}^{I\gamma}\right| \leqslant C\big(n, |I|, |I_1|, |I_2|\big)\, t^{-|I_1|}. \tag{3.2.2}$$

Proof. Consider the identity

$$[Z^I, \underline{\partial}_\beta]u = [Z^I, \Phi_\beta^\gamma \partial_\gamma]u = \sum_{\substack{I_1+I_2=I\\|I_2|<|I|}} Z^{I_1}\Phi_\beta^\gamma Z^{I_2}\partial_\gamma u + \Phi_\beta^\gamma [Z^I, \partial_\gamma]u.$$

In the first sum, we commute Z^{I_2} and ∂_γ and obtain

$$\begin{aligned}
[Z^I, \underline{\partial}_\beta]u &= \sum_{\substack{I_1+I_2=I\\|I_2|<|I|}} Z^{I_1}\Phi_\beta^\gamma \partial_\gamma Z^{I_2} u + \sum_{\substack{I_1+I_2=I\\|I_2|<|I|}} Z^{I_1}\Phi_\beta^\gamma [Z^{I_2}, \partial_\gamma]u + \Phi_\beta^\gamma [Z^I, \partial_\gamma]u \\
&= \sum_{\substack{I_1+I_2=I\\|I_2|<|I|}} Z^{I_1}\Phi_\beta^\gamma \partial_\gamma Z^{I_2} u + \sum_{I_1+I_2=I} Z^{I_1}\Phi_\beta^\gamma [Z^{I_2}, \partial_\gamma]u \\
&= \sum_{\substack{I_1+I_2=I\\|I_2|<|I|}} Z^{I_1}\Phi_\beta^\gamma \partial_\gamma Z^{I_2} u + \sum_{\substack{I_1+I_2=I\\|J|<|I_2|}} (Z^{I_1}\Phi_\beta^\gamma)\theta_{\gamma J}^{I_2\alpha}\partial_\alpha Z^J u.
\end{aligned}$$

Hence, $\underline{\theta}_{\gamma J}^{I\alpha}$ are linear combinations of $Z^{I_1}\Phi_\beta^\gamma$ and $\theta_{\gamma J}^{I_2\alpha} Z^{I_1}\Phi_\beta^\gamma$ with $|I_1| \leqslant |I|$ and $|I_2| \leqslant |I|$, which yields (3.2.1). Note that $\theta_{\gamma J}^{I_2\alpha}$ are constants, so that

$$\partial^{I_3} Z^{I_4}\big(\theta_{\gamma J}^{I_2\alpha} Z^{I_1}\Phi_\beta^\gamma\big) = \theta_{\gamma J}^{I_2\alpha}\partial^{I_3} Z^{I_4} Z^{I_1}\Phi_\beta^\gamma$$

and, by (2.2.11), we arrive at (3.2.2). □

Lemma 3.2.2 (Algebraic decompositions of commutators. III). *For all sufficiently regular functions u defined in the cone $\mathcal{K}$, the following identities hold:*

$$[Z^I, \underline{\partial}_c]u = \sum_{|J|<|I|} \sigma_{cJ}^{Ib}\underline{\partial}_b Z^J u + t^{-1}\sum_{|J'|<|I|} \rho_{cJ'}^{I\gamma}\partial_\gamma Z^{J'} u, \tag{3.2.3}$$

where the functions $\sigma_{cJ}^{I\gamma}$ and $\rho_{cJ}^{I\gamma}$ are smooth and satisfy in $\mathcal{K}$:

$$\left|\partial^{I_1} Z^{I_2}\sigma_{\beta J}^{I\gamma}\right| \leqslant C(n, |I|, |I_1|, |I_2|)t^{-|I_1|}, \tag{3.2.4a}$$

$$\left|\partial^{I_1} Z^{I_2}\rho_{\beta J}^{I\gamma}\right| \leqslant C(n, |I|, |I_1|, |I_2|)t^{-|I_1|}. \tag{3.2.4b}$$

Proof. We proceed by induction. The case $|I| = 1$ is easily checked from (3.1.2b), (3.1.2c) and (3.1.4). Assuming that (3.2.3) with (3.2.4a) and (3.2.4b) hold for $|I| \leqslant k$, we consider indices $|I| = k+1$. To do this, let Z^I be a product of operators with index $|I| = k+1$. Then, $Z^I = Z_1 Z^{I'}$ with $|I'| = k$ and the following holds:

$$[Z^I, \underline{\partial}_c]u = [Z_1 Z^{I'}, \underline{\partial}_c] = Z_1\big([Z^{I'}, \underline{\partial}_c]u\big) + [Z_1, \underline{\partial}_c]Z^{I'}u.$$

Case $Z_1 = \partial_\alpha$. In this case, we write

$$\begin{aligned}
&Z_1\big([Z^{I'}, \underline{\partial}_c]u\big) \\
&= \partial_\alpha\Bigg(\sum_{|J|<|I'|} \sigma_{cJ}^{I'b}\underline{\partial}_b Z^J u + t^{-1}\sum_{|J'|<|I'|} \rho_{cJ'}^{I'\gamma}\partial_\gamma Z^{J'}u\Bigg) \\
&= \sum_{|J|<|I'|} \partial_\alpha\big(\sigma_{cJ}^{I'b}\big)\underline{\partial}_b Z^J u + \sum_{|J|<|I'|} \sigma_{cJ}^{I'b}\underline{\partial}_b\partial_\alpha Z^J u + \sum_{|J|<|I'|} \sigma_{cJ}^{I'b}[\partial_\alpha, \underline{\partial}_b]Z^J u \\
&\quad + \partial_\alpha t^{-1}\sum_{|J'|<|I'|} \big(\rho_{cJ'}^{I'\gamma}\big)\partial_\gamma Z^{J'}u + t^{-1}\sum_{|J'|<|I'|} \partial_\alpha\big(\rho_{cJ'}^{I'\gamma}\big)\partial_\gamma Z^{J'}u \\
&\quad + t^{-1}\sum_{|J'|<|I'|} \rho_{cJ'}^{I'\gamma}\partial_\gamma\partial_\alpha Z^{J'}u.
\end{aligned}$$

For the third term, we recall (3.1.2b) and write

$$\sigma_{cJ}^{I'b}[\partial_\alpha, \underline{\partial}_b]Z^J u = t^{-1}\sigma_{cJ}^{I'b}\underline{\Gamma}_{\alpha b}^{\gamma}\partial_\gamma Z^J u.$$

Hence, we find

$$\begin{aligned}
&Z_1\big([Z^{I'}, \underline{\partial}_c]u\big) \\
&= \sum_{|J|<|I'|} \partial_\alpha\big(\sigma_{cJ}^{I'b}\big)\underline{\partial}_b Z^J u + \sum_{|J|<|I'|} \sigma_{cJ}^{I'b}\underline{\partial}_b\partial_\alpha Z^J u + t^{-1}\sum_{|J|<|I'|} \sigma_{cJ}^{I'b}\underline{\Gamma}_{\alpha b}^{\gamma}\partial_\gamma Z^I u \\
&\quad + \partial_\alpha t^{-1}\sum_{|J'|<|I'|} \big(\rho_{cJ'}^{I'\gamma}\big)\partial_\gamma Z^{J'}u + t^{-1}\sum_{|J'|<|I'|} \partial_\alpha\big(\rho_{cJ'}^{I'\gamma}\big)\partial_\gamma Z^{J'}u \\
&\quad + t^{-1}\sum_{|J'|<|I'|} \rho_{cJ'}^{I'\gamma}\partial_\gamma\partial_\alpha Z^{J'}u.
\end{aligned}$$

We conclude that, when $Z_1 = \partial_\alpha$, $\sigma_{\alpha J}^{I\gamma}$ are linear combinations of $\partial_\alpha\sigma_{cJ}^{I'b}$ and $\sigma_{cJ}^{I'b}$ with $|J| < |I'|$. For all $\partial^{I_1}Z^{I_2}$, we have

$$\begin{aligned}
\partial^{I_1}Z^{I_2}\big(\partial_\alpha\sigma_{cJ}^{I'b}\big) &= \partial_\alpha\partial^{I_1}Z^{I_2}\sigma_{cJ}^{I'b} + \partial^{I_1}\big([Z^{I_2}, \partial_\alpha]\sigma_{cJ}^{I'b}\big) \\
&= \partial_\alpha\partial^{I_1}Z^{I_2}\sigma_{cJ}^{I'b} + \sum_{|I_2'|<|I_2|} \theta_{\alpha I_2'}^{I_2\beta}\partial_\beta\partial^{I_1}Z^{I_2'}\sigma_{cJ}^{I_2\beta}.
\end{aligned}$$

By applying the induction assumption, we find

$$\left|\partial_\alpha \partial^{I_1} Z^{I_2} \sigma_{cJ}^{I'b}\right| \leqslant C(n, I_1, I_2, I') t^{-|I_1|-1},$$

so that

$$\left|\partial^{I_1} Z^{I_2} \big(\partial_\alpha \sigma_{cJ}^{I'b}\big)\right| \leqslant C(n, I_1, I_2, I') t^{-|I_1|}.$$

Similarly, $\sigma_{cJ}^{I'b}$ satisfies the same estimate and we see that, in this case, $\sigma_{\alpha J}^{I\gamma}$ satisfies the desired estimate.

Note that $\rho_{\alpha J}^{I\gamma}$ are linear combinations of the following terms:

$$\sigma_{cJ}^{I'b} \underline{\Gamma}_{\alpha b}^{\gamma}, \quad t\partial_\alpha t^{-1} \rho_{cJ}^{I'\gamma}, \quad \partial_\alpha \rho_{cJ}^{I\gamma}, \quad \rho_{cJ}^{I'\gamma}.$$

Observe that $\underline{\Gamma}_{\alpha\beta}^{\gamma}$ are linear combinations of Φ_α^β and recall the estimate (2.2.11), we see that each term can be controlled by $C(n, I_1, I_2, I) t^{-|I_1|}$.

Case $Z_1 = L_a$**.** In this case, we write

$$
\begin{aligned}
& Z_1\big([Z^{I'}, \underline{\partial}_c] u\big) \\
&= L_a\Big(\sum_{|J|<|I'|} \sigma_{cJ}^{I'b} \underline{\partial}_b Z^J u + t^{-1} \sum_{|J'|<|I'|} \rho_{cJ'}^{I'\gamma} \partial_\gamma Z^{J'} u\Big) \\
&= \sum_{|J|<|I'|} \big(L_a \sigma_{cJ}^{I'b}\big) \underline{\partial}_b Z^J u + \sum_{|J|<|I'|} \sigma_{cJ}^{I'b} \underline{\partial}_b L_a Z^J u + \sum_{|J|<|I'|} \sigma_{cJ}^{I'b} [L_a, \underline{\partial}_b] Z^J u \\
&\quad + t^{-1} (t L_a t^{-1}) \sum_{\substack{\gamma \\ |J'|<|I'|}} \rho_{cJ'}^{I'\gamma} \partial_\gamma Z^{J'} u \\
&\quad + t^{-1} \sum_{|J'|<|I'|} \big(L_a \rho_{cJ'}^{I'\gamma}\big) \partial_\gamma Z^{J'} u + t^{-1} \sum_{|J'|<|I'|} \rho_{cJ'}^{I'\gamma} \partial_\gamma L_a Z^{J'} u \\
&\quad + t^{-1} \sum_{|J'|<|I'|} \rho_{cJ'}^{I'\gamma} [L_a, \partial_\gamma] Z^{J'} u.
\end{aligned}
$$

In the right-hand side, we apply (3.1.2c) on the third term and (3.1.2a) on the last term, and observe that the coefficient of the fourth term

$$t L_a (t^{-1}) = -\frac{x^a}{t} = \Psi_a^0.$$

We have

$$
\begin{aligned}
& Z_1\big([Z^{I'}, \underline{\partial}_c] u\big) \\
&= \sum_{|J|<|I'|} \big(L_a \sigma_{cJ}^{I'b}\big) \underline{\partial}_b Z^J u + \sum_{|J|<|I'|} \sigma_{cJ}^{I'b} \underline{\partial}_b L_a Z^J u + \sum_{|J|<|I'|} \sigma_{cJ}^{I'b} \underline{\Theta}_{ab}^{c} \underline{\partial}_c Z^J u \\
&\quad + t^{-1} \Psi_a^0 \sum_{|J'|<|I'|} \rho_{cJ'}^{I'\gamma} \partial_\gamma Z^{J'} u \\
&\quad + t^{-1} \sum_{|J'|<|I'|} \big(L_a \rho_{cJ'}^{I'\gamma}\big) \partial_\gamma Z^{J'} u + t^{-1} \sum_{|J'|<|I'|} \rho_{cJ'}^{I'\gamma} \partial_\gamma L_a Z^{J'} u \\
&\quad + t^{-1} \sum_{|J'|<|I'|} \rho_{cJ'}^{I'\gamma} \Theta_{a\gamma}^{\beta} \partial_\beta Z^{J'} u.
\end{aligned}
$$

We also recall that

$$[Z_1, \underline{\partial}_c] Z^{I'} u = [L_a, \partial_c] Z^{I'} u = \underline{\Theta}_{ac}^b \underline{\partial}_b Z^{I'} u.$$

We observe that σ_{cJ}^{Ib} are linear combinations of $L_a \sigma_{cJ}^{I'b}$, $\sigma_{cJ}^{I'b}$, $\sigma_{cJ}^{I'b} \underline{\Theta}_{ab}^c$ and $\underline{\Theta}_{ac}^b$. The estimate on the first two terms is a direct consequence of the induction assumption (3.2.4a). On the other hand, the third term is estimated as follows:

$$\partial^{I_1} Z^{I_2} \sigma_{cJ}^{I'b} \underline{\Theta}_{ab}^c = \sum_{\substack{I_3+I_4=I_1 \\ I_5+I_6=I_2}} \partial^{I_3} Z^{I_5} \sigma_{cJ}^{I'b} \, \partial^{I_4} Z^{I_6} \underline{\Theta}_{ab}^c.$$

The first factor is bounded by the induction assumption (3.2.4a), and the second factor can be bounded by (2.2.12), by observing that $\underline{\Theta}_{ab}^c$ is a linear combination of Ψ_γ^δ. In the same manner, we get the desired estimate on $\underline{\Theta}_{ac}^b$. So, (3.2.4a) is established in the case $Z_1 = L_a$, $|I| = k + 1$.

In the same way, $\rho_{cJ}^{I\gamma}$ are linear combinations of $\Psi_a^0 \rho_{cJ'}^{I'\gamma}$, $L_a \rho_{cJ'}^{I'\gamma}$, $\rho_{cJ'}^{I'\gamma}$, $\rho_{cJ'}^{I'\gamma} \Theta_{a\gamma}^\beta$. The second and the third terms are bounded by the induction assumption (3.2.4b). The first term is estimated by applying (2.2.12) and the induction assumption (3.2.4b). For the last term, note that $\Theta_{a\gamma}^\beta$ are linear combinations of $\Psi_\delta^{\delta'}$ and constants. Hence, by applying (2.2.12) and the induction assumption (3.2.4b), the desired bound is reached. Finally, by combining these estimates together, (3.2.4b) is established in the case $|I| = k + 1$, which completes the argument and, consequently, the proof of Lemma 3.2.1. □

3.3 Estimates of commutators

The following statement is now immediate in view of (3.1.5), (3.2.1), and (3.2.3).

Lemma 3.3.1. *For all sufficiently regular functions u defined in the cone $\mathcal{K}$, the following estimates hold:*

$$\big|[Z^I, \underline{\partial}_\alpha] u\big| + \big|[Z^I, \partial_\alpha] u\big| \leqslant C(n, |I|) \sum_{\beta, |J|<I} \big|\partial_\beta Z^J u\big|, \tag{3.3.1}$$

$$\big|[Z^I, \underline{\partial}_a] u\big| \leqslant C(n, |I|) \sum_{b, |J_1|<|I|} \big|\underline{\partial}_b Z^{J_1} u\big| + C(n, |I|) \sum_{\gamma, J_2|<|I|} \big|t^{-1} \partial_\gamma Z^{J_2} u\big|. \tag{3.3.2}$$

Furthermore, from (3.1.5), (3.2.1), and (3.2.3), we deduce the following estimates. Recall our convention $Z^I = 0$ in case we write $|I| < 0$.

Lemma 3.3.2. *For all sufficiently regular functions u defined in the cone $\mathcal{K}$, the following estimates hold:*

$$\big|[Z^I, \partial_\alpha \partial_\beta] u\big| \leqslant C(n, |I|) \sum_{\substack{\gamma, \gamma' \\ |J| < |I|}} \big|\partial_\gamma \partial_{\gamma'} Z^J u\big|, \tag{3.3.3}$$

$$\begin{aligned} \big|[Z^I, \underline{\partial}_a \underline{\partial}_\beta] u\big| + \big|[Z^I, \underline{\partial}_\alpha \underline{\partial}_b] u\big| \leqslant & C(n, |I|) \sum_{\substack{c, \gamma \\ |J_1| \leqslant |I|}} \big|\underline{\partial}_c \underline{\partial}_\gamma Z^{J_1} u\big| \\ & + C(n, |I|) t^{-1} \sum_{\substack{\gamma \\ |J_2| \leqslant |I|}} \big|\partial_\gamma Z^{J_2} u\big|. \end{aligned} \tag{3.3.4}$$

Proof. 1. To derive (3.3.3), we write

$$[Z^I, \partial_\alpha \partial_\beta] u = \partial_\alpha \big([Z^I, \partial_\beta] u\big) + [Z^I, \partial_\alpha] \partial_\beta u$$

and, by applying (3.1.5), the first term in the right-hand side can be written as

$$\partial_\alpha \big([Z^I, \partial_\beta] u\big) = \partial_\alpha \Big(\sum_{|J| < |I|} \theta_{\beta J}^{I\gamma} \partial_\gamma Z^J u \Big) = \sum_{|J| < |I|} \theta_{\beta J}^{I\gamma} \partial_\alpha \partial_\gamma Z^J u,$$

which is bounded by $C(n, |I|) \sum_{\substack{\alpha, \beta \\ |J| < |I|}} \big|\partial_\alpha \partial_\beta Z^J u\big|$. The second term is estimated as follows:

$$\begin{aligned} [Z^I, \partial_\alpha] \partial_\beta u &= \sum_{|J| < |I|} \theta_{\alpha J}^{I\gamma} \partial_\gamma Z^J \partial_\beta u \\ &= \sum_{|J| < |I|} \theta_{\alpha J}^{I\gamma} \partial_\gamma \partial_\beta Z^J u + \sum_{|J| < |I|} \theta_{\alpha J}^{I\gamma} \partial_\gamma [Z^J, \partial_\beta] u \\ &= \sum_{|J| < |I|} \theta_{\alpha J}^{I\gamma} \partial_\gamma \partial_\beta Z^J u + \sum_{|J| < |I|} \theta_{\alpha J}^{I\gamma} \partial_\gamma \Big(\sum_{|J'| < |J|} \theta_{\beta J'}^{J\delta} \partial_\delta Z^{J'} u \Big) \\ &= \sum_{|J| < |I|} \theta_{\alpha J}^{I\gamma} \partial_\gamma \partial_\beta Z^J u + \sum_{\substack{|J| < |I| \\ |J'| < |J|}} \theta_{\alpha J}^{I\gamma} \theta_{\beta J'}^{J\delta} \partial_\gamma \partial_\delta Z^{J'} u. \end{aligned}$$

This latter expression is bounded by $C(n, |I|) \sum_{\substack{\alpha, \beta \\ |J| < |I|}} \big|\partial_\alpha \partial_\beta Z^J u\big|$, and (3.3.3) is established.

2. We now derive (3.3.4) and we begin by considering $[Z^I, \underline{\partial}_a\underline{\partial}_\beta]u$. By (3.2.1) and (3.2.3), we have

$$
\begin{aligned}
&[Z^I, \underline{\partial}_a\underline{\partial}_\beta]u \\
&= \underline{\partial}_a[Z^I, \underline{\partial}_\beta]u + [Z^I, \underline{\partial}_a]\underline{\partial}_\beta u \\
&= \underline{\partial}_a\Big(\sum_{|J|<|I|} \underline{\theta}_{\beta J}^{I\gamma}\partial_\gamma Z^J u\Big) + \sum_{|J|<|I|} \sigma_{aJ}^{Ic}\underline{\partial}_c Z^J \underline{\partial}_\beta u + t^{-1} \sum_{|J|<|I|} \rho_{aJ}^{I\gamma}\partial_\gamma Z^J \underline{\partial}_\beta u \\
&= \sum_{|J|<|I|} \underline{\partial}_a\underline{\theta}_{\beta J}^{I\gamma}\partial_\gamma Z^J u + \sum_{|J|<|I|} \underline{\theta}_{\beta J}^{I\gamma}\underline{\partial}_a\partial_\gamma Z^J u \\
&\quad + \sum_{|J|<|I|} \sigma_{aJ}^{Ic}\underline{\partial}_c Z^J \underline{\partial}_\beta u + t^{-1} \sum_{|J|<|I|} \rho_{aJ}^{I\gamma}\partial_\gamma Z^J \underline{\partial}_\beta u,
\end{aligned}
$$

thus

$$
\begin{aligned}
&[Z^I, \underline{\partial}_a\underline{\partial}_\beta]u \\
&= \sum_{|J|<|I|} \underline{\partial}_a\underline{\theta}_{\beta J}^{I\gamma}\partial_\gamma Z^J u + \sum_{|J|<|I|} \underline{\theta}_{\beta J}^{I\gamma}\underline{\partial}_a\partial_\gamma Z^J u \\
&\quad + \sum_{|J|<|I|} \sigma_{aJ}^{Ic}\underline{\partial}_c\underline{\partial}_\beta Z^J u + \sum_{|J|<|I|} \sigma_{aJ}^{Ic}\underline{\partial}_c\big([Z^J, \underline{\partial}_\beta]u\big) + t^{-1} \sum_{|J|<|I|} \rho_{aJ}^{I\gamma}\partial_\gamma Z^J \underline{\partial}_\beta u
\end{aligned}
$$

and, therefore,

$$
\begin{aligned}
&[Z^I, \underline{\partial}_a\underline{\partial}_\beta]u \\
&= \sum_{|J|<|I|} \underline{\partial}_a\underline{\theta}_{\beta J}^{I\gamma}\partial_\gamma Z^J u + \sum_{|J|<|I|} \underline{\theta}_{\beta J}^{I\gamma}\underline{\partial}_a\big(\Psi_\gamma^{\gamma'}\underline{\partial}_{\gamma'} Z^J u\big) \\
&\quad + \sum_{|J|<|I|} \sigma_{aJ}^{Ic}\underline{\partial}_c\underline{\partial}_\beta Z^J u + \sum_{|J|<|I|} \sigma_{aJ}^{Ic}\underline{\partial}_c\big([Z^J, \underline{\partial}_\beta]u\big) + t^{-1} \sum_{|J|<|I|} \rho_{aJ}^{I\gamma}\partial_\gamma Z^J \underline{\partial}_\beta u \\
&= \sum_{|J|<|I|} \underline{\theta}_{\beta J}^{I\gamma}\Psi_\gamma^{\gamma'}\underline{\partial}_a\underline{\partial}_{\gamma'} Z^J u + \sum_{|J|<|I|} \sigma_{aJ}^{Ic}\underline{\partial}_c\underline{\partial}_\beta Z^J u \\
&\quad + \sum_{|J|<|I|} \underline{\theta}_{\beta J}^{I\gamma}\underline{\partial}_a\big(\Psi_\gamma^{\gamma'}\big)\underline{\partial}_{\gamma'} Z^J u + \sum_{|J|<|I|} \underline{\partial}_a\underline{\theta}_{\beta J}^{I\gamma}\partial_\gamma Z^J u \\
&\quad + \sum_{|J|<|I|} \sigma_{aJ}^{Ic}\underline{\partial}_c\big([Z^J, \underline{\partial}_\beta]u\big) + t^{-1} \sum_{|J|<|I|} \rho_{aJ}^{I\gamma}\partial_\gamma Z^J \underline{\partial}_\beta u.
\end{aligned}
\tag{3.3.5}
$$

Here, the first and second terms can be bounded by $\sum_{a,\beta,|J|<|I|} |\underline{\partial}_a\underline{\partial}_\beta Z^J u|$. By recalling that $|\underline{\partial}_a\Psi_\gamma^{\gamma'}| \leqslant C(n)t^{-1}$ and $|\underline{\partial}_a\underline{\theta}_{\beta J}^{I\gamma}| \leqslant C(n,|I|)t^{-1}$, the third and fourth can be bounded by $C(n,|I|)t^{-1}\sum_{\gamma,|J_2|\leqslant|I|} |\partial_\gamma Z^{J_2} u|$.

So, we focus on the fifth term:

$$\begin{aligned}
&\sum_{|J|<|I|} \sigma_{aJ}^{Ic}\underline{\partial}_c([Z^J,\underline{\partial}_\beta]u)\\
&= \sum_{|J|<|I|} \sigma_{aJ}^{Ic}\underline{\partial}_c\Big(\sum_{|J'|<|J|} \underline{\theta}_{\beta J'}^{J\gamma}\partial_\gamma u\Big)\\
&= \sum_{\substack{|J|<|I|\\|J'|<|J|}} \sigma_{aJ}^{Ic}\underline{\partial}_c(\underline{\theta}_{\beta J'}^{J\gamma})\partial_\gamma u + \sum_{\substack{|J|<|I|\\|J'|<|J|}} \sigma_{aJ}^{Ic}\underline{\theta}_{\beta J'}^{J\gamma}\underline{\partial}_c\partial_\gamma u\\
&= \sum_{\substack{|J|<|I|\\|J'|<|J|}} \sigma_{aJ}^{Ic}\underline{\partial}_c(\underline{\theta}_{\beta J'}^{J\gamma})\partial_\gamma u + \sum_{\substack{|J|<|I|\\|J'|<|J|}} \sigma_{aJ}^{Ic}\underline{\theta}_{\beta J'}^{J\gamma}\underline{\partial}_c(\Psi_\gamma^{\gamma'}\underline{\partial}_{\gamma'}u),
\end{aligned}$$

so

$$\begin{aligned}
&\sum_{|J|<|I|} \sigma_{\beta J}^{Ic}\underline{\partial}_c([Z^J,\underline{\partial}_\beta]u)\\
&= \sum_{\substack{|J|<|I|\\|J'|<|J|}} \sigma_{aJ}^{Ic}\underline{\partial}_c(\underline{\theta}_{\beta J'}^{J\gamma})\partial_\gamma u + \sum_{\substack{|J|<|I|\\|J'|<|J|}} \sigma_{aJ}^{Ic}\underline{\theta}_{\beta J'}^{J\gamma}\underline{\partial}_c(\Psi_\gamma^{\gamma'})\underline{\partial}_{\gamma'}u\\
&\quad + \sum_{\substack{|J|<|I|\\|J'|<|J|}} \sigma_{aJ}^{Ic}\underline{\theta}_{\beta J'}^{J\gamma}\Psi_\gamma^{\gamma'}\underline{\partial}_c\underline{\partial}_{\gamma'}u.
\end{aligned}$$

Similarly, the first two terms can be bounded by

$$C(n,|I|)t^{-1}\sum_{\gamma,|J_2|\leqslant|I|} |\partial_\gamma Z^{J_2}u|$$

and the last term is bounded by $\sum_{\alpha\beta,|J|<|I|} |\underline{\partial}_a\underline{\partial}_\beta Z^J u|$.

For the last term in (3.3.5), we perform the following calculation:

$$\begin{aligned}
&t^{-1}\sum_{|J|<|I|} \rho_{aJ}^{I\gamma}\partial_\gamma Z^J\underline{\partial}_\beta u\\
&= t^{-1}\sum_{|J|<|I|} \rho_{aJ}^{I\gamma}\partial_\gamma\underline{\partial}_\beta Z^J u + t^{-1}\sum_{|J|<|I|} \rho_{aJ}^{I\gamma}\partial_\gamma([Z^J,\underline{\partial}_\beta]u)\\
&= t^{-1}\sum_{|J|<|I|} \rho_{aJ}^{I\gamma}\partial_\gamma(\Psi_\beta^{\beta'}\partial_{\beta'}Z^J u) + t^{-1}\sum_{|J|<|I|} \rho_{aJ}^{I\gamma}\partial_\gamma(\underline{\theta}_{\beta J'}^{J\delta}\partial_\delta Z^{J'}u)\\
&= t^{-1}\sum_{|J|<|I|} \rho_{aJ}^{I\gamma}\Psi_\beta^{\beta'}\partial_\gamma\partial_{\beta'}Z^J u + t^{-1}\sum_{|J|<|I|} \rho_{aJ}^{I\gamma}\partial_\gamma\Psi_\beta^{\beta'}\partial_{\beta'}Z^J u\\
&\quad + t^{-1}\sum_{|J|<|I|} \rho_{aJ}^{I\gamma}\underline{\theta}_{\beta J'}^{J\delta}\partial_\gamma\partial_\delta Z^{J'}u + t^{-1}\sum_{|J|<|I|} \rho_{aJ}^{I\gamma}\partial_\gamma\underline{\theta}_{\beta J'}^{J\delta}\partial_\delta Z^{J'}u.
\end{aligned}$$

The desired estimates on these terms are immediate thanks to (3.2.4b), (3.2.2), and the fact that $|\partial_\alpha\Psi_\beta^{\beta'}| \leqslant Ct^{-1}$.

Now, in order to control $[Z^I, \underline{\partial}_\alpha \underline{\partial}_b]u$, we consider the identity

$$[Z^I, \underline{\partial}_\alpha \underline{\partial}_b]u = [Z^I, \underline{\partial}_b \underline{\partial}_\alpha]u + [Z^I, [\underline{\partial}_\alpha, \underline{\partial}_b]]u.$$

The estimate on the first term in the right-hand side is already done. We concentrate on the second term, via the following calculation:

$$\begin{aligned}
&[Z^I, [\underline{\partial}_\alpha, \underline{\partial}_b]]u \\
&= [Z^I, \Phi_\alpha^\beta [\partial_\beta, \underline{\partial}_b]]u - [Z^I, \underline{\partial}_b \Phi_\alpha^\beta \partial_\beta]u \\
&= [Z^I, t^{-1}\Phi_\alpha^\beta \underline{\Gamma}_{\beta b}^\gamma \partial_\gamma]u - [Z^I, \underline{\partial}_b \Phi_\alpha^\beta \partial_\beta]u \\
&= \sum_{\substack{I_1+I_2=I \\ |I_2|<|I|}} Z^{I_1}\big(t^{-1}\Phi_\alpha^\beta \underline{\Gamma}_{\beta b}^\gamma\big) Z^{I_2}\partial_\gamma u + t^{-1}\Phi_\alpha^\beta \underline{\Gamma}_{\beta b}^\gamma [Z^I, \partial_\gamma]u \\
&\quad - \sum_{\substack{I_1+I_2=I \\ |I_2|<|I|}} Z^{I_1}\big(\underline{\partial}_b \Phi_\alpha^\beta\big) Z^{I_2}\partial_\beta u - \underline{\partial}_b \phi_\alpha^\beta [Z^I, \partial_\beta]u
\end{aligned}$$

and thus

$$\begin{aligned}
&[Z^I, [\underline{\partial}_\alpha, \underline{\partial}_b]]u \\
&= \sum_{\substack{I_1+I_2=I \\ |I_2|<|I|}} Z^{I_1}\big(t^{-1}\Phi_\alpha^\beta \underline{\Gamma}_{\beta b}^\gamma\big)\partial_\gamma Z^{I_2} u \\
&\quad + \sum_{\substack{I_1+I_2=I \\ |I_2|<|I|}} Z^{I_1}\big(t^{-1}\Phi_\alpha^\beta \underline{\Gamma}_{\beta b}^\gamma\big)[Z^{I_2}, \partial_\gamma]u + t^{-1}\Phi_\alpha^\beta \underline{\Gamma}_{\beta b}^\gamma [Z^I, \partial_\gamma]u \\
&\quad - \sum_{\substack{I_1+I_2=I \\ |I_2|<|I|}} Z^{I_1}\big(\underline{\partial}_b \Phi_\alpha^\beta\big)\partial_\beta Z^{I_2} u \\
&\quad - \sum_{\substack{I_1+I_2=I \\ |I_2|<|I|}} Z^{I_1}\big(\underline{\partial}_b \Phi_\alpha^\beta\big)[Z^{I_2}, \partial_\beta]u - \underline{\partial}_b \phi_\alpha^\beta [Z^I, \partial_\beta]u.
\end{aligned}$$

So, we have

$$\begin{aligned}
&[Z^I, [\underline{\partial}_\alpha, \underline{\partial}_b]]u \\
&= \sum_{\substack{I_1+I_2=I \\ |I_2|<|I|}} Z^{I_1}\big(t^{-1}\Phi_\alpha^\beta \underline{\Gamma}_{\beta b}^\gamma\big)\partial_\gamma Z^{I_2} u + \sum_{I_1+I_2=I} Z^{I_1}\big(t^{-1}\Phi_\alpha^\beta \underline{\Gamma}_{\beta b}^\gamma\big)[Z^{I_2}, \partial_\gamma]u \\
&\quad - \sum_{\substack{I_1+I_2=I \\ |I_2|<|I|}} Z^{I_1}\big(\underline{\partial}_b \Phi_\alpha^\beta\big)\partial_\beta Z^{I_2} u - \sum_{I_1+I_2=I} Z^{I_1}\big(\underline{\partial}_b \Phi_\alpha^\beta\big)[Z^{I_2}, \partial_\beta]u.
\end{aligned}$$

We observe that the function $t^{-1}\Phi_\alpha^\beta \underline{\Gamma}_{\beta b}^\gamma$ and $\underline{\partial}_b \Phi_\alpha^\beta$ are homogeneous of degree -1 and are smooth in the closed region $\{t \geqslant 1, |x| \leqslant t\}$.

By Lemma 2.2.1, we have

$$\begin{aligned}
\big|Z^{I_1}\big(t^{-1}\Phi_\alpha^\beta \underline{\Gamma}_{\beta b}^\gamma\big)\big| &\leqslant C(n, |I|)t^{-1}, \\
\big|Z^{I_1}\big(\underline{\partial}_b \Phi_\alpha^\beta\big)\big| &\leqslant C(n, |I|)t^{-1}.
\end{aligned}$$

We conclude with

$$\big|[Z^I, [\underline{\partial}_\alpha, \underline{\partial}_b]]u\big| \leqslant C(n, |I|)t^{-1} \sum_{\substack{\gamma \\ |J|<|I|}} \big|\partial_\gamma Z^J u\big|,$$

and this completes the proof of (3.3.4). □

Lemma 3.3.3. *For all sufficiently regular functions u defined in the cone $\mathcal{K}$, the following estimates hold:*

$$\big|Z^I\big((s/t)\partial_\alpha u\big)\big| \leqslant \big|(s/t)\partial_\alpha Z^I u\big| + C(n, |I|) \sum_{\beta, |J|<|I|} \big|(s/t)\partial_\beta Z^J u\big|. \quad (3.3.6)$$

The proof of this lemma will rely on the following technical remark.

Lemma 3.3.4. *For all index I, the function*

$$\Xi^I := (t/s) Z^I (s/t) \quad (3.3.7)$$

defined in the closed cone $\overline{\mathcal{K}} = \{|x| \leqslant t-1\}$, is smooth and all of its derivatives (of any order) are bounded in $\overline{\mathcal{K}}$. Furthermore, it is homogeneous of degree η with $\eta \leqslant 0$.

We admit here this result and give the proof of (3.3.6), while the proof of (3.3.7) is given afterwards.

Proof of Lemma 3.3.3. We observe that

$$\begin{aligned}
[Z^I, (s/t)\partial_\alpha]u &= \sum_{\substack{I_1+I_2=I \\ |I_1|<|I|}} Z^{I_2}(s/t)\, Z^{I_1}\partial_\alpha u + (s/t)[Z^I, \partial_\alpha]u \\
&= \sum_{\substack{I_1+I_2=I \\ |I_1|<|I|}} Z^{I_2}(s/t)\, [Z^{I_1}, \partial_\alpha]u + \sum_{\substack{I_1+I_2=I \\ |I_1|<|I|}} Z^{I_2}(s/t)\, \partial_\alpha Z^{I_1} u \\
&\quad + (s/t)[Z^I, \partial_\alpha]u \\
&= \sum_{I_1+I_2=I} Z^{I_2}(s/t)\, [Z^{I_1}, \partial_\alpha]u + \sum_{\substack{I_1+I_2=I \\ |I_1|<|I|}} Z^{I_2}(s/t)\, \partial_\alpha Z^{I_1}.
\end{aligned}$$

By applying (3.3.7), we find

$$\begin{aligned}
\Big|\sum_{I_1+I_2=I} Z^{I_2}(s/t)\, [Z^{I_1}, \partial_\alpha]u\Big| &\leqslant \sum_{I_1+I_2=I} \big|Z^{I_2}(s/t)\big|\, \big|[Z^{I_1}, \partial_\alpha]u\big| \\
&\leqslant C(n, |I|) \sum_{|J|<|I|} \big|(s/t)\partial_\alpha Z^J u\big|
\end{aligned}$$

and

$$\Big|\sum_{\substack{I_1+I_2=I \\ |I_1|<|I|}} Z^{I_2}(s/t)\, \partial_\alpha Z^{I_1} u\Big| \leqslant C(n, |I|) \sum_{|J|<|I|} \big|(s/t)\partial_\alpha Z^J u\big|.$$

□

Proof of Lemma 3.3.4. We consider the identities

$$\begin{aligned}&(t/s)L_a(s/t) = -x^a/t,\\&(t/s)\partial_a(s/t) = -x^a/s^2, \quad (t/s)\partial_t(s/t) = |x|^2s^{-2}t^{-1}.\end{aligned} \tag{3.3.8}$$

In the cone $\mathcal{K} = \{|x| < t-1\}$, the functions x^a/s^2, $|x|^2s^{-2}t^{-1}$, and x^a/t are smooth and bounded, while x^a/s^2 and $|x|^2s^{-2}t^{-1}$ are homogeneous of degree -1, and x^a/t is homogeneous of degree 0. All of their derivatives (of any order) are bounded in $\overline{\mathcal{K}}$. We have proved (3.3.7) in the case $|I| = 1$.

For the case $|I| > 1$, we use an induction on $|I|$. Assume that for $|I| \leqslant k$, (3.3.7) holds. For an operator Z^I with $|I| = k+1$, we suppose that $Z^I = Z_1 Z^{I'}$ where $|I'| = k$ and we have

$$(t/s)Z^I(s/t) = (t/s)Z_1Z^{I'}(s/t) = Z_1\big((t/s)Z^{I'}(s/t)\big) - Z_1(s/t)Z^{I'}(s/t),$$

where Z_1 can be ∂_α or L_a. By the induction assumption and (3.3.8), the second term is a smooth function and is homogeneous of non-positive degree, while all of its derivatives are bounded in $\overline{\mathcal{K}}$. We focus on the first term and, by the induction assumption, $(t/s)Z^{I'}(s/t)$ is smooth, homogeneous of non-positive degree, and is bounded in $\overline{\mathcal{K}}$. We need to distinguish between two different cases, as follows.

Case $Z_1 = \partial_\alpha$. In this case, $Z_1\big((t/s)Z^{I'}(s/t)\big)$ is homogeneous of degree less than or equal to -1, and by the induction assumption, all of its derivatives are bounded in $\overline{\mathcal{K}}$.

Case $Z_1 = L_a$. In this case, $Z_1\big((t/s)Z^{I'}(s/t)\big)$ is homogeneous of degree less than or equal to 0.

$$Z_1\big((t/s)Z^{I'}(s/t)\big) = x^a\partial_t\big((t/s)Z^{I'}(s/t)\big) + t\partial_a\big((t/s)Z^{I'}(s/t)\big).$$

Denote by $f(t,x) = \partial_\alpha\big((t/s)Z^{I'}(s/t)\big)$ and recall that $f(t,x)$ is homogeneous of degree η with $\eta \leqslant -1$ so

$$f(t,x) = (t/2)^\eta f(2, 2x/t).$$

Recall that $f(2,x)$ is bounded when $|x| \leqslant 2$. Recalling that $t \geqslant 1$, we get

$$\big|\partial_\alpha\big((t/s)Z^{I'}(s/t)\big)\big| = |f(t,x)| \leqslant C(n,I')t^\eta \leqslant C(n,I')t^{-1}$$

thus

$$\big|Z_1\big((t/s)Z^{I'}(s/t)\big)\big| \leqslant C(n,I')(1+|x^a|/t).$$

Taking into consideration the fact that in $\overline{\mathcal{K}}$, $|x^a| \leqslant t$, the desired result is proven. □

Chapter 4

The null structure in the semi-hyperboidal frame

4.1 Estimating first-order derivatives

In this chapter, we derive various estimates on null quadratic or cubic forms. Recall that a quadratic form $T^{\alpha\beta}\partial_\alpha u \partial_\beta v$ acting on the gradient of functions u, v is said to satisfy the **null condition** if, for all null vectors $\xi \in \mathbb{R}^4$, i.e. all vectors satisfying $(\xi_0)^2 - \sum_a (\xi_a)^2 = 0$, one has

$$T^{\alpha\beta}\xi_\alpha\xi_\beta = 0. \tag{4.1.1}$$

Similarly, a cubic form $A^{\alpha\beta\gamma}\xi_\alpha\xi_\beta\xi_\gamma$ is said to satisfy the null condition if, for all null vectors, one has

$$A^{\alpha\beta\gamma}\xi_\alpha\xi_\beta\xi_\gamma = 0. \tag{4.1.2}$$

All the terms of interest will be linear combinations of factors $\partial_\alpha u \, \partial_\beta v$, $\partial_\gamma u \partial_\alpha \partial_\beta v$, and $u \partial_\alpha \partial_\beta v$. Throughout, the notation $s^2 = t^2 - r^2$ is in order.

Proposition 4.1.1. *If $T^{\alpha\beta}\xi_\alpha\xi_\beta$ is a quadratic form satisfying the null condition, then for every index I there exists a constant $C(I) > 0$ such that*

$$\left|Z^I \underline{T}^{00}\right| \leqslant C(I)\,\frac{t^2 - r^2}{t^2} = C(I)\,(s/t)^2 \quad \text{in the cone } \mathcal{K}. \tag{4.1.3}$$

Similarly, if a cubic form $A^{\alpha\beta\gamma}\xi_\alpha\xi_\beta\xi_\gamma$ satisfies the null condition, then for every index I there exists a constant $C(I) > 0$ such that

$$\left|Z^I \underline{A}^{000}\right| \leqslant C(I)\,\frac{t^2 - r^2}{t^2} = C(I)\,(s/t)^2 \quad \text{in the cone } \mathcal{K}. \tag{4.1.4}$$

The proof will be based on a homogeneity lemma stated now, which concerns homogeneous functions defined "near" the light cone in $\{r = t-1\}$.

Lemma 4.1.1 (Homogeneity lemma). *Let f be a smooth function defined in the closed set $\{t \geqslant 1\} \cap \{t/2 \leqslant r \leqslant t\}$. Assume that f is homogeneous of degree η in the sense that*

$$f(pt, px) = p^\eta f(1, x/t), \qquad t/2 \leqslant |x| \leqslant t, \quad pt \geqslant 1, \quad t \geqslant 1.$$

The following estimate holds:

$$\left|Z^I f(t,x)\right| \leqslant C(n, |I|) t^\eta \qquad \text{in the region } \{r \geqslant t/2\} \cap \mathcal{K}. \tag{4.1.5}$$

The proof is similar to the one of Lemma 2.2.1 and is omitted.

Proof of Proposition 4.1.1. We use the notation $-\omega_a = \omega^a = x^a/|x|$ and $\omega_0 = \omega^0 = 1$, so that $(\omega_0)^2 - \sum_a (\omega_a)^2 = 0$. We consider the component $\underline{T}^{00}$ and write

$$\begin{aligned}\underline{T}^{00} = T^{\alpha\beta}\Psi_\alpha^0\Psi_\beta^0 &= T^{\alpha\beta}\Psi_\alpha^0\Psi_\beta^0 - T^{\alpha\beta}\omega_\alpha\omega_\beta \\ &= T^{\alpha\beta}\big(\Psi_\alpha^0\Psi_\beta^0 - \omega_\alpha\omega_\beta\big).\end{aligned}$$

First, we consider the region "away" from the light cone. When $r \leqslant t/2$, we have

$$Z^I\big(T^{\alpha\beta}\Psi_\alpha^0\Psi_\beta^0\big) = \sum_{I_1+I_2=I} T^{\alpha\beta} Z^{I_1}\Psi_\alpha^0 Z^{I_2}\Psi_\beta^0.$$

Applying Lemma 2.2.1, we obtain

$$\left|Z^I\big(T^{\alpha\beta}\Psi_\alpha^0\Psi_\beta^0\big)\right| \leqslant \sum_{I_1+I_2=I} K\left|Z^{I_1}\Psi_\alpha^0\right|\left|Z^{I_2}\Psi_\beta^0\right| \leqslant C(n,|I|)K(s/t)^2(t/s)^2$$

for some constant $K > 0$. Recall that when $0 \leqslant r \leqslant t/2$, we have $(t/s)^2 \leqslant 4/3$ so in the region $\{r \leqslant t/2\} \cap \mathcal{K}$,

$$\left|Z^I\big(T^{\alpha\beta}\Psi_\alpha^0\Psi_\beta^0\big)\right| \leqslant C(n,|I|)K(s/t)^2.$$

Second, in the region $\{r \geqslant t/2\} \cap \mathcal{K}$, we have

$$\underline{T}^{00} = T^{\alpha\beta}\big(\Psi_\alpha^0\Psi_\beta^0 - \omega_\alpha\omega_\beta\big)$$

and, thus,

$$Z^I\underline{T}^{00} = T^{\alpha\beta}Z^I\big(\Psi_\alpha^0\Psi_\beta^0 - \omega_\alpha\omega_\beta\big).$$

When $\alpha = \beta = 0$, we have $\big(\Psi_\alpha^0\Psi_\beta^0 - \omega_\alpha\omega_\beta\big) = 0$. When $\alpha = a > 0, \beta = 0$, we have

$$\begin{aligned}Z^I\big(\Psi_\alpha^0\Psi_\beta^0 - \omega_\alpha\omega_\beta\big) &= -Z^I\big(\omega_a(1-(r/t))\big) \\ &= -\sum_{I_1+I_2=I} Z^{I_1}\omega_a\, Z^{I_2}\left(\frac{t-r}{t}\right).\end{aligned}$$

When $\alpha = a > 0$, $\beta = 0 > 0$, we obtain

$$Z^I\left(\Psi_\alpha^0\Psi_\beta^0 - \omega_\alpha\omega_\beta\right) = Z^I\left(\frac{x^a x^b}{t^2} - \frac{x^a x^b}{r^2}\right)$$
$$= \sum_{I_1+I_2+I_3+I_4=I} Z^{I_1}\omega_a\, Z^{I_2}\omega_b\, Z^{I_3}\left(1+\frac{r}{t}\right) Z^{I_4}\left(\frac{r-t}{t}\right).$$

We focus on the estimates of $Z^I\omega_a$, $Z^I\left(1+\frac{r}{t}\right)$ and $Z^I\left(\frac{t-r}{t}\right)$. By Lemma 4.1.1, $Z^I\omega_a$ and $Z^I\left(1+\frac{r}{t}\right)$ are bounded by $C(n,|I|)$. For the estimate of $Z^I\left(\frac{t-r}{t}\right)$, we write

$$\frac{t-r}{t} = \frac{s^2}{t^2}\frac{t}{t+r}$$

and then

$$Z^I\left((t-r)/t\right) = \sum_{I_1+I_2+I_3=I} Z^{I_1}\left(t/(t+r)\right)Z^{I_2}(s/t)Z^{I_3}(s/t).$$

We observe that $t/(t+r)$ is smooth in $\{t \geqslant 1\} \cap \{t/2 \leqslant r \leqslant t\}$ and is homogeneous of degree 0. We have

$$\left|Z^{I_1}\left(t/(t+r)\right)\right| \leqslant C(|I|)$$

and the term $Z^{I_2}(s/t)$ is bounded by $C(|I|)(s/t)$, thanks to Lemma 3.3.4. We conclude with

$$\left|Z^I\left((t/r)/t\right)\right| \leqslant C(|I|)(s/t)^2.$$

□

Next, we have the following result concerning null quadratic forms.

Proposition 4.1.2 (Estimate of null forms). *For all null quadratic form $T^{\alpha\beta}\partial_\alpha u\partial_\beta v$ with constant coefficients $T^{\alpha\beta}$ and for any index I, one has*

$$\left|Z^I\left(T^{\alpha\beta}\partial_\alpha u\partial_\beta v\right)\right| \leqslant CK(s/t)^2 \sum_{|I_1|+|I_2|\leqslant|I|} \left|Z^{I_1}\partial_t u Z^{I_2}\partial_t v\right|$$
$$+ CK \sum_{\substack{a,\beta,\\ |I_1|+|I_2|\leqslant|I|}} \left(\left|Z^{I_1}\underline{\partial}_a u\, Z^{I_2}\underline{\partial}_\beta v\right| + \left|Z^{I_1}\underline{\partial}_\beta u\, Z^{I_2}\underline{\partial}_a v\right|\right),$$

with $K = \max_{\alpha,\beta}\left|T^{\alpha\beta}\right|$.

The importance of this estimate lies on the factor $(s/t)^2$ in front of the component $Z^{I_1}\partial_t u Z^{I_2}\partial_t v$. As we will see later, the derivative $\underline{\partial}_a$ enjoys better L^∞ and L^2 estimates in our framework. The derivatives of direction ∂_t do not always have enough decay, and the factor $(s/t)^2$ precisely allows us to overcome this potential lack of L^2 and L^∞ decay.

Proof. The proof is based on a change of frame. In the semi-hyperboloidal frame we have

$$
\begin{aligned}
T^{\alpha\beta}\partial_\alpha u \partial_\beta v &= \underline{T}^{\alpha\beta}\underline{\partial}_\alpha u \underline{\partial}_\beta v \\
&= \underline{T}^{00}\partial_t u \partial_t v + \underline{T}^{0b}\partial_t u \underline{\partial}_b v + \underline{T}^{a0}\underline{\partial}_a u \partial_t v + \underline{T}^{ab}\underline{\partial}_a u \underline{\partial}_b v.
\end{aligned}
$$

We have

$$
\begin{aligned}
& Z^I\left(T^{\alpha\beta}\partial_\alpha u \partial_\beta v\right) \\
&= Z^I\left(\underline{T}^{\alpha\beta}\underline{\partial}_\alpha u \underline{\partial}_\beta v\right) \\
&= Z^I\left(\underline{T}^{00}\partial_t u \partial_t v\right) + Z^I\left(\underline{T}^{0b}\partial_t u \underline{\partial}_b v\right) + Z^I\left(\underline{T}^{a0}\underline{\partial}_a u \partial_t v\right) + Z^I\left(\underline{T}^{ab}\underline{\partial}_a u \underline{\partial}_b v\right) \\
&=: R_1 + R_2 + R_3 + R_4.
\end{aligned}
$$

Recalling the null condition satisfied by the null form under consideration and by Proposition 4.1.1, we get

$$
\begin{aligned}
|R_1| \leqslant & \sum_{I_1+I_2+I_3=I} \left|Z^{I_3}\left(\underline{T}^{00}\right) Z^{I_1}\left(\partial_t u\right) Z^{I_2}\left(\partial_t v\right)\right| \\
\leqslant & CK(s/t)^2 \sum_{|I_1|+|I_2|\leqslant|I|} \left|Z^{I_1}\partial_t u\right| \left|Z^{I_2}\partial_t v\right|.
\end{aligned}
$$

The term R_2 are estimated directly. Recalling that by Lemma 2.2.2, $|Z^I \underline{T}^{\alpha\beta}| \leqslant C(I)K$, we find

$$
\begin{aligned}
|R_2| \leqslant & \sum_{\substack{b \\ I_1+I_2+I_3=I}} \left|Z^{I_3}\left(\underline{T}^{0b}\right) Z^{I_1}\left(\partial_t u\right) Z^{I_2}\left(\underline{\partial}_b v\right)\right| \\
\leqslant & CK \sum_{\substack{b \\ |I_1|+|I_2|\leqslant|I|}} \left|Z^{I_1}\left(\partial_t u\right) Z^{I_2}\left(\underline{\partial}_b v\right)\right|.
\end{aligned}
$$

The terms R_3 and R_4 are estimated similarly and the proof is complete. □

4.2 Estimating second-order derivatives

We can also deal with second-order derivatives $B^{\alpha\beta}\partial_\alpha\partial_\beta u$, with constant $B^{\alpha\beta}$. Recall that a second-order operator is said to satisfy the null condition if

$$
B^{\alpha\beta}\xi_\alpha\xi_\beta = 0 \quad \text{when } \xi_0\xi_0 - \sum_{i=1}^{3}\xi_i\xi_i = 0.
$$

Proposition 4.2.1. *Let $B^{\alpha\beta}\partial_\alpha\partial_\beta$ be a second-order operator satisfying the null condition with constants $B^{\alpha\beta}$ bounded by K. One has*

$$\begin{aligned}&\left|Z^I\left(B^{\alpha\beta}\partial_\alpha\partial_\beta u\right)\right| \\ &\leqslant C(n,|I|)K(s/t)^2\sum_{|I_1|\leqslant|I|}Z^I\left(\partial_t\partial_t u\right)+C(n,|I|)K\sum_{\substack{a,\alpha\\|I_1|\leqslant|I|}}Z^I\left(\underline{\partial}_a\partial_\alpha u\right)\\ &\quad + C(n,|I|)\frac{K}{t}\sum_{\alpha,|I_1|\leqslant|I|}\left|Z^{I_1}\left(\partial_\alpha u\right)\right|.\end{aligned}$$

Proof. We have

$$\begin{aligned}B^{\alpha\beta}\partial_\alpha\partial_\beta u =&\underline{B}^{\alpha\beta}\underline{\partial}_\alpha\underline{\partial}_\beta u+B^{\alpha\beta}\partial_\alpha\left(\Psi_\beta^{\beta'}\right)\underline{\partial}_{\beta'}u\\ =&\underline{B}^{00}\partial_t\partial_t u+\underline{B}^{a0}\underline{\partial}_a\partial_t u+\underline{B}^{0b}\partial_t\underline{\partial}_b u+\underline{B}^{ab}\underline{\partial}_a\underline{\partial}_b u\\ &+B^{\alpha\beta}\partial_\alpha\left(\Psi_\beta^{\beta'}\right)\underline{\partial}_{\beta'}u.\end{aligned}$$

Recall that, for a null quadratic form, $|Z^J\underline{B}^{00}|\leqslant C(n,|J|)K(s/t)^2$ and

$$\begin{aligned}\left|Z^I\underline{B}^{00}\partial_t\partial_t u\right|\leqslant&\sum_{I_1+I_2=I}\left|Z^{I_2}\underline{B}^{00}\right|\left|Z^{I_1}\partial_t\partial_t u\right|\\ \leqslant&\, C(n,|I|)K(s/t)^2\sum_{|I_1|\leqslant|I|}\left|Z^{I_1}\partial_t\partial_t u\right|.\end{aligned}$$

Also, from Lemma 2.2.2, we have $\left|Z^I\underline{B}\right|\leqslant C(n,|I|)|B|$ and

$$\begin{aligned}\left|Z^I\left(\underline{B}^{a0}\underline{\partial}_a\partial_t u\right)\right|\leqslant&\sum_{\substack{a\\I_1+I_2=I}}\left|Z^{I_2}\underline{B}^{a0}\right|\left|Z^{I_1}\underline{\partial}_a\partial_t u\right|\\ \leqslant&\, C(n,|I|)K\sum_{\substack{a\\|I_1|\leqslant|I|}}\left|Z^{I_1}\underline{\partial}_a\partial_t u\right|.\end{aligned}$$

For the term $Z^I\left(\underline{B}^{0b}\partial_t\underline{\partial}_b u\right)$ by applying (3.1.2) we have

$$\begin{aligned}\left|Z^I\left(\underline{B}^{0b}\partial_t\underline{\partial}_b u\right)\right|\leqslant&\sum_{I_1+I_2=I}\left|Z^{I_2}\underline{B}^{b0}\right|\left|Z^{I_1}\partial_t\underline{\partial}_b u\right|\\ \leqslant&\, C(n,|I|)K\sum_{\substack{b\\|I_1|\leqslant|I|}}\left|Z^{I_1}\partial_t\underline{\partial}_b u\right|\end{aligned}$$

and

$$\begin{aligned}\left|Z^{I_1}\partial_t\underline{\partial}_b u\right|\leqslant&\left|Z^{I_1}\underline{\partial}_b\partial_t u\right|+\left|Z^{I_1}[\underline{\partial}_b,\partial_t]u\right|\\ \leqslant&\left|Z^{I_1}\underline{\partial}_b\partial_t u\right|+\left|Z^{I_1}\left(t^{-1}\underline{\Gamma}_{0b}^\gamma\partial_\gamma u\right)\right|\\ \leqslant&\left|Z^{I_1}\underline{\partial}_b\partial_t u\right|+\sum_{I_3+I_4=I_2}\left|Z^{I_4}\left(t^{-1}\underline{\Gamma}_{0b}^\gamma\right)Z^{I_3}\partial_\gamma u\right)\right|\\ \leqslant&\left|Z^{I_1}\underline{\partial}_b\partial_t u\right|+Ct^{-1}\sum_{\substack{\gamma\\|I_3|\leqslant|I_1|}}\left|Z^{I_3}\partial_\gamma u\right|.\end{aligned}$$

Here, we have used the estimate $\big|Z^J\big(t^{-1}\underline{\Gamma}_{0b}^{\gamma}\big)\big| \leqslant C(n,|J|)t^{-1}$ because $t^{-1}\underline{\Gamma}_{0b}^{\gamma}$ is homogeneous of degree -1. So we have proven that

$$\big|Z^{I_1}\underline{B}^{0b}\partial_t\underline{\partial}_b u\big| \leqslant C(n,|I|)K\sum_{\substack{b\\|I_1|\leqslant|I|}}\big|Z^{I_1}\underline{\partial}_b\partial_t u\big| + C(n,|I|)Kt^{-1}\sum_{\substack{\gamma\\|I_1|\leqslant|I|}}\big|Z^{I_1}\partial_\gamma u\big|.$$

The term $Z^I\big(\underline{B}^{ab}\underline{\partial}_a\underline{\partial}_b u\big)$ is estimated as follows:

$$\begin{aligned}\big|Z^I\big(\underline{B}^{ab}\underline{\partial}_a\underline{\partial}_b u\big)\big| &\leqslant C(n,|I|)\sum_{\substack{a,b\\I_1+I_2=I}}\big|Z^{I_2}\underline{B}^{ab}\big|\,\big|Z^{I_1}\underline{\partial}_a\underline{\partial}_b u\big|\\ &\leqslant C(n,|I|)K\sum_{\substack{\beta,a\\I_1+I_2=I}}\big|Z^{I_1}\underline{\partial}_a\underline{\partial}_\beta u\big|.\end{aligned}$$

The term $Z^I\big(B^{\alpha\beta}\partial_\alpha\big(\Psi_\beta^{\beta'}\big)\underline{\partial}_{\beta'}u\big)$ is estimated by applying the decay rate supplied by $Z^J\big|\partial_\alpha\big(\Psi_\beta^{\beta'}\big)\big| \leqslant C(n,|J|)t^{-1}$. This completes the proof. □

As a direct corollary, the following estimates hold.

Proposition 4.2.2. *Consider a bilinear form $B^{\alpha\beta}u\partial_\alpha\partial_\beta v$ acting on the function u and the Hessian of v and suppose that the quadratic form $B^{\alpha\beta}\partial_\alpha\partial_\beta$ satisfies the null condition. The following estimates hold:*

$$\begin{aligned}\big|Z^I\big(B^{\alpha\beta}u\partial_\alpha\partial_\beta v\big)\big| \leqslant{}& C(n,|I|)K(s/t)^2\sum_{|I_1|+|I_2|\leqslant|I|}\big|Z^{I_1}u\big|\,\big|Z^{I_2}\partial_t\partial_t v\big|\\ &+C(n,|I|)K\sum_{\substack{\alpha,b\\|I_1|+|I_2|\leqslant|I|}}\big|Z^{I_1}u\big|\,\big|Z^{I_2}\underline{\partial}_\alpha\underline{\partial}_b v\big|\\ &+C(n,|I|)K\sum_{\substack{a,\beta\\|I_1|+|I_2|\leqslant|I|}}\big|Z^{I_1}u\big|\,\big|Z^{I_2}\underline{\partial}_a\underline{\partial}_\beta v\big|\\ &+C(n,|I|)Kt^{-1}\sum_{\substack{\alpha\\|I_1|+|I_2|\leqslant|I|}}\big|Z^{I_1}u\big|\,\big|Z^{I_2}\underline{\partial}_\alpha v\big|,\end{aligned}\tag{4.2.1a}$$

$$\begin{aligned}\big|[Z^I,B^{\alpha\beta}u\partial_\alpha\partial_\beta]v\big| \leqslant{}& C(n,|I|)K(s/t)^2\sum_{\substack{|I_1|+|I_2|\leqslant|I|\\|I_2|<|I|}}\big|Z^{I_1}u\big|\,\big|Z^{I_2}\partial_t\partial_t v\big|\\ &+C(n,|I|)K\sum_{a,\beta}\sum_{\substack{|I_1|+|I_2|\leqslant|I|\\|I_2|<|I|}}\big|Z^{I_1}u\big|\,\big|Z^{I_2}\underline{\partial}_a\underline{\partial}_\beta v\big|\\ &+C(n,|I|)K\sum_{\alpha,b}\sum_{\substack{|I_1|+|I_2|\leqslant|I|\\|I_2|<|I|}}\big|Z^{I_1}u\big|\,\big|Z^{I_2}\underline{\partial}_\alpha\underline{\partial}_b v\big|\\ &+C(n,|I|)Kt^{-1}\sum_{\substack{|I_1|+|I_2|\leqslant|I|\\|I_2|<|I|,\beta}}\big|Z^{I_1}u\big|\,\big|Z^{I_2}\partial_{\beta'}v\big|.\end{aligned}\tag{4.2.1b}$$

Proof. We first establish (4.2.1a):

$$
\begin{aligned}
Z^I\big(B^{\alpha\beta}u\partial_\alpha\partial_\beta v\big) =& Z^I\big(\underline{B}^{\alpha\beta}u\underline{\partial}_\alpha\underline{\partial}_\beta v + B^{\alpha\beta}u\partial_\alpha\Psi_\beta^{\beta'}\underline{\partial}_{\beta'}v\big)\\
=& \sum_{I_1+I_2+I_3=I} Z^{I_3}\underline{B}^{\alpha\beta}\,Z^{I_1}u\,Z^{I_2}\underline{\partial}_\alpha\underline{\partial}_\beta v\\
&+ \sum_{I_1+I_2+I_3=I} B^{\alpha\beta}Z^{I_1}u\,Z^{I_3}\partial_\alpha\Psi_\beta^{\beta'}\,Z^{I_2}\underline{\partial}_{\beta'}v =: R_1 + R_2,
\end{aligned}
$$

in which R_1 can be estimated as follows:

$$
\begin{aligned}
R_1 =& \sum_{I_1+I_2+I_3=I} Z^{I_3}\underline{B}^{\alpha\beta}\,Z^{I_1}u\,Z^{I_2}\underline{\partial}_\alpha\underline{\partial}_\beta v\\
=& \sum_{I_1+I_2+I_3=I} Z^{I_3}\underline{B}^{00}\,Z^{I_1}u\,Z^{I_2}\partial_t\partial_t v\\
&+ \sum_{I_1+I_2+I_3=I} Z^{I_3}\underline{B}^{a0}\,Z^{I_1}u\,Z^{I_2}\underline{\partial}_a\partial_t v\\
&+ \sum_{I_1+I_2+I_3=I} Z^{I_3}\underline{B}^{0b}\,Z^{I_1}u\,Z^{I_2}\partial_t\underline{\partial}_b v\\
&+ \sum_{I_1+I_2+I_3=I} Z^{I_3}\underline{B}^{ab}\,Z^{I_1}u\,Z^{I_2}\underline{\partial}_a\underline{\partial}_b v.
\end{aligned}
$$

Since $\big|Z^I\underline{B}^{\alpha\beta}\big| \leqslant C(n,|I|)K$ and $\big|Z^I\underline{B}^{00}\big| \leqslant C(n,|I|)(s/t)^2$, we have

$$
\begin{aligned}
|R_1| \leqslant& C(n,|I|)K(s/t)^2 \sum_{|I_1|+|I_2|\leqslant|I|} \big|Z^{I_1}u\big|\,\big|Z^{I_2}\partial_t\partial_t v\big|\\
&+ C(n,|I|)K \sum_{\substack{\alpha,b\\|I_1|+|I_2|\leqslant|I|}} \big|Z^{I_1}u\big|\,\big|Z^{I_2}\underline{\partial}_\alpha\underline{\partial}_b v\big|\\
&+ C(n,|I|)K \sum_{\substack{a,\beta\\|I_1|+|I_2|\leqslant|I|}} \big|Z^{I_1}u\big|\,\big|Z^{I_2}\underline{\partial}_a\underline{\partial}_\beta v\big|.
\end{aligned}
$$

Observe now that $\big|Z^I\partial_\alpha\Psi_\beta^{\beta'}\big| \leqslant C(n,|I|)t^{-1}$, then R_2 is bounded by

$$
\begin{aligned}
|R_2| \leqslant& \sum_{I_1+I_2+I_3=I} \big|B^{\alpha\beta}\big|\,\big|Z^{I_1}u\big|\,\big|Z^{I_3}\partial_\alpha\Psi_\beta^{\beta'}\big|\,\big|Z^{I_2}\underline{\partial}_{\beta'}v\big|\\
\leqslant& C(n,|I|)Kt^{-1} \sum_{\substack{\alpha\\|I_1|+|I_2|\leqslant|I|}} \big|Z^{I_1}u\big|\,\big|Z^{I_2}\underline{\partial}_\alpha v\big|.
\end{aligned}
$$

In view of $Z^{I_2}\underline{\partial}_\alpha v = Z^{I_2}\big(\Psi_\alpha^{\alpha'}\partial_{\alpha'}v\big) = Z^{I_4}\Psi_\alpha^{\alpha'}Z^{I_5}\partial_{\alpha'}v$ and that $Z^{I_4}\Psi_\alpha^{\alpha'}$ is bounded, the estimate (4.2.1a) is proven.

Next, in order to derive (4.2.1b), we observe that

$$
[Z^I, B^{\alpha\beta}u\partial_\alpha\partial_\beta]v = [Z^I, \underline{B}^{\alpha\beta}u\underline{\partial}_\alpha\underline{\partial}_\beta]v + [Z^I, B^{\alpha\beta}u\partial_\alpha\Psi_\beta^{\beta'}\underline{\partial}_{\beta'}]v =: R_3 + R_4.
$$

The term R_3 is decomposed as follows:

$$
\begin{aligned}
&[Z^I, \underline{B}^{\alpha\beta} u \underline{\partial}_\alpha \underline{\partial}_\beta] v \\
&= \sum_{\substack{I_1+I_2+I_3 \\ |I_2|<|I|}} Z^{I_3} \underline{B}^{\alpha\beta}\, Z^{I_1} u\, Z^{I_2} \underline{\partial}_\alpha \underline{\partial}_\beta v + \underline{B}^{\alpha\beta} u [Z^I, \underline{\partial}_\alpha \underline{\partial}_\beta] v \\
&=: R_5 + R_6.
\end{aligned}
$$

The term R_5 is estimated as follows:

$$
\begin{aligned}
R_5 = & \sum_{\substack{I_1+I_2+I_3 \\ |I_2|<|I|}} Z^{I_3} \underline{B}^{\alpha\beta}\, Z^{I_1} u\, Z^{I_2} \underline{\partial}_\alpha \underline{\partial}_\beta v \\
= & \sum_{\substack{I_1+I_2+I_3 \\ |I_2|<|I|}} Z^{I_3} \underline{B}^{00}\, Z^{I_1} u\, Z^{I_2} \partial_t \partial_t v \\
& + \sum_{\substack{I_1+I_2+I_3 \\ |I_2|<|I|}} Z^{I_3} \underline{B}^{a0}\, Z^{I_1} u\, Z^{I_2} \underline{\partial}_a \partial_t v + \sum_{\substack{I_1+I_2+I_3 \\ |I_2|<|I|}} Z^{I_3} \underline{B}^{0b}\, Z^{I_1} u\, Z^{I_2} \partial_t \underline{\partial}_b v \\
& + \sum_{\substack{I_1+I_2+I_3 \\ |I_2|<|I|}} Z^{I_3} \underline{B}^{ab}\, Z^{I_1} u\, Z^{I_2} \underline{\partial}_a \underline{\partial}_b v,
\end{aligned}
$$

thus

$$
\begin{aligned}
|R_5| \leqslant & \sum_{\substack{I_1+I_2+I_3=I \\ |I_2|<|I|}} \big|Z^{I_3} \underline{B}^{00}\big|\, \big|Z^{I_1} u\big|\, \big|Z^{I_2} \partial_t \partial_t v\big| \\
& + \sum_{\substack{I_1+I_2+I_3=I \\ |I_2|<|I|}} \big|Z^{I_3} \underline{B}^{a0}\big|\, \big|Z^{I_1} u\big|\, \big|Z^{I_2} \underline{\partial}_a \partial_t v\big| \\
& + \sum_{\substack{I_1+I_2+I_3=I \\ |I_2|<|I|}} \big|Z^{I_3} \underline{B}^{0b}\big|\, \big|Z^{I_1} u\big|\, \big|Z^{I_2} \partial_t \underline{\partial}_b v\big| \\
& + \sum_{\substack{I_1+I_2+I_3=I \\ |I_2|<|I|}} \big|Z^{I_3} \underline{B}^{ab}\big|\, \big|Z^{I_1} u\big|\, \big|Z^{I_2} \underline{\partial}_a \underline{\partial}_b v\big|,
\end{aligned}
$$

hence

$$
\begin{aligned}
|R_5| \leqslant & C(n, |I|) K (s/t)^2 \sum_{\substack{|I_1|+|I_2| \leqslant |I| \\ |I_2|<|I|}} \big|Z^{I_1} u\big|\, \big|Z^{I_2} \partial_t \partial_t v\big| \\
& + C(n, |I|) K \sum_{\substack{|I_1|+|I_2| \leqslant |I| \\ |I_2|<|I|}} \big|Z^{I_1} u\big|\, \big|Z^{I_2} \underline{\partial}_a \underline{\partial}_\beta v\big| \\
& + C(n, |I|) K \sum_{\substack{|I_1|+|I_2| \leqslant |I| \\ |I_2|<|I|}} \big|Z^{I_1} u\big|\, \big|Z^{I_2} \underline{\partial}_\alpha \underline{\partial}_b v\big|.
\end{aligned}
$$

The term R_6 is decomposed as follows:

$$\underline{B}^{\alpha\beta}u[Z^I,\underline{\partial}_\alpha\underline{\partial}_\beta]v = \underline{B}^{00}u[Z^I,\partial_t\partial_t]v + \underline{B}^{a0}u[Z^I,\underline{\partial}_a\partial_t]v + \underline{B}^{0b}u[Z^I,\partial_t\underline{\partial}_b]v + \underline{B}^{ab}u[Z^I,\underline{\partial}_a\underline{\partial}_b]v.$$

We apply (3.3.3) and (3.3.4) and that $|\underline{B}^{00}| \leqslant C(s/t)^2$:

$$\begin{aligned}
&|\underline{B}^{\alpha\beta}u[Z^I,\underline{\partial}_\alpha\underline{\partial}_\beta]v| \\
&= |\underline{B}^{00}u[Z^I,\partial_t\partial_t]v| + |\underline{B}^{a0}u[Z^I,\underline{\partial}_a\partial_t]v| \\
&\quad + |\underline{B}^{0b}u[Z^I,\partial_t\underline{\partial}_b]v| + |\underline{B}^{ab}u[Z^I,\underline{\partial}_a\underline{\partial}_b]v| \\
&\leqslant C(n,|I|)K(s/t)^2|u| \sum_{\substack{\gamma,\gamma' \\ |I'|<|I|}} |Z^{I_1}\partial_\gamma\partial_{\gamma'}v| \\
&\quad + C(n,|I|)K|u| \sum_{\substack{a,\beta \\ |I_1|<|I|}} |\underline{\partial}_a\underline{\partial}_\beta Z^{I_1}v| + C(n,|I|)Kt^{-1}|u| \sum_{\substack{\gamma \\ |I_1|<|I|}} |\partial_\gamma Z^{I_1}v|.
\end{aligned}$$

The term R_4 is estimated as follows:

$$\begin{aligned}
&[Z^I, B^{\alpha\beta}u\partial_\alpha\Psi_\beta^{\beta'}\underline{\partial}_{\beta'}]v \\
&= \sum_{\substack{I_1+I_2+I_3=I \\ |I_2|<|I|}} B^{\alpha\beta}Z^{I_1}u\, Z^{I_3}\partial_\alpha\Psi_\beta^{\beta'}\, Z^{I_2}\underline{\partial}_{\beta'}v + B^{\alpha\beta}u\partial_\alpha\Psi_\beta^{\beta'}[Z^I,\underline{\partial}_{\beta'}]v.
\end{aligned}$$

Thanks to the additional decreasing rate in $|Z^I\partial_\alpha\Psi_\beta^{\beta'}| \leqslant C(n,|I|)t^{-1}$, the first term is bounded by

$$C(n,|I|)Kt^{-1} \sum_{\substack{|I_1|+|I_2|\leqslant|I| \\ |I_2|<|I|,\beta'}} |Z^{I_1}u|\,|Z^{I_2}\partial_{\beta'}v|.$$

The second term is estimated by (3.3.2) and the additional decreasing rate supplied by $|Z^I\partial_\alpha\Psi_\beta^{\beta'}| \leqslant C(n,|I|)t^{-1}$. It can also be bounded by

$$C(n,|I|)Kt^{-1} \sum_{\substack{|I_1|+|I_2|\leqslant|I| \\ |I_2|<|I|,\beta}} |Z^{I_1}u|\,|Z^{I_2}\partial_{\beta'}v|.$$

This completes the proof of (4.2.1b). □

4.3 Products of first-order and second-order derivatives

The third type of null form we treat is a null quadratic form acting on the gradient and the Hessian, as now stated.

Proposition 4.3.1. *Consider a quadratic form acting on the gradient of u and the Hessian of v, that is, $A^{\alpha\beta\gamma}\partial_\gamma u\partial_\alpha\partial_\beta v$, for functions u, v defined* *in the cone $\mathcal{K}$, and suppose that $A^{\alpha\beta\gamma}$ satisfies the null condition. The following estimate holds for any index I:*

$$\begin{aligned}
&\big|Z^I\big(A^{\alpha\beta\gamma}\partial_\gamma u\partial_\alpha\partial_\beta v\big)\big| \\
&\leqslant C(n,|I|)K(s/t)^2 \sum_{|I_1|+|I_2|\leqslant|I|} \big|Z^{I_1}\partial_t u\, Z^{I_2}\partial_t\partial_t v\big| \\
&\quad + C(n,|I|)K\big(\Omega_1(I,u,v)+\Omega_2(I,u,v)\big),
\end{aligned}$$

where

$$\begin{aligned}
\Omega_1(I,u,v) = &\sum_{\substack{a,\beta,\gamma\\ |I_1|+|I_2|\leqslant|I|}} \big|Z^{I_1}\underline{\partial}_a u\big|\,\big|Z^{I_2}\underline{\partial}_\beta\underline{\partial}_\gamma v\big| + \sum_{\substack{\alpha,b,\gamma\\ |I_1|+|I_2|\leqslant|I|}} \big|Z^{I_1}\underline{\partial}_\alpha u\big|\,\big|Z^{I_2}\underline{\partial}_b\underline{\partial}_\gamma v\big| \\
&+ \sum_{\substack{\alpha\beta,c\\ |I_1|+|I_2|\leqslant|I|}} \big|Z^{I_1}\underline{\partial}_\alpha u\big|\,\big|Z^{I_2}\underline{\partial}_\beta\underline{\partial}_c v\big|
\end{aligned}$$

and

$$\Omega_2(I,u,v) \leqslant t^{-1} \sum_{\substack{\alpha,\beta,\\ |I_1|+|I_2|\leqslant|I|}} \big|Z^{I_1}\partial_\alpha u\, Z^{I_2}\partial_\beta v\big|.$$

Proof. We observe the following change of frame formula:

$$\begin{aligned}
A^{\alpha\beta\gamma}\partial_\gamma u\partial_\alpha\partial_\beta v &= \underline{A}^{\alpha\beta\gamma}\underline{\partial}_\gamma u\underline{\partial}_\alpha\underline{\partial}_\beta v + A^{\alpha\beta\gamma}\partial_\gamma u\partial_\alpha\big(\Psi_\beta^{\beta'}\big)\underline{\partial}_{\beta'}u \\
&=: R_1(u,v)+R_2(u,v).
\end{aligned}$$

The term $Z^I R_2$ can be estimated as follows. Recall that $|Z^I\partial_\alpha\big(\Psi_\beta^{\beta'}\big)| \leqslant \frac{C}{t}$, then

$$Z^I\big(A^{\alpha\beta\gamma}\partial_\gamma u\partial_\alpha\big(\Psi_\beta^{\beta'}\big)\underline{\partial}_{\beta'}v\big) = \sum_{I_1+I_2+I_3=I} A^{\alpha\beta\gamma}Z^{I_1}\partial_\gamma u\, Z^{I_2}\underline{\partial}_{\beta'}v\, Z^{I_3}\partial_\alpha\big(\Psi_\beta^{\beta'}\big),$$

which can be estimated as

$$\big|Z^I\big(A^{\alpha\beta\gamma}\partial_\gamma u\partial_\alpha\big(\Psi_\beta^{\beta'}\big)\underline{\partial}_{\beta'}v\big)\big| \leqslant CKt^{-1} \sum_{\substack{\alpha,\beta\\ |I_1|+|I_2|\leqslant|I|}} \big|Z^{I_1}\partial_\alpha u Z^{I_2}\partial_\beta v\big| = \Omega_2.$$

The term R_1 can be estimated as follows:

$$\begin{aligned}
R_1 =& \underline{A}^{\alpha\beta\gamma}\underline{\partial}_\gamma u\underline{\partial}_\alpha\underline{\partial}_\beta v \\
=& \underline{A}^{000}\partial_t u\partial_t\partial_t v \\
&+ \underline{A}^{00c}\underline{\partial}_c u\partial_t\partial_t v + \underline{A}^{0b0}\partial_t u\partial_t\underline{\partial}_b v + \underline{A}^{a00}\partial_t u\underline{\partial}_a\partial_t v \\
&+ \underline{A}^{ab0}\partial_t u\underline{\partial}_a\underline{\partial}_b v + \underline{A}^{0bc}\underline{\partial}_c u\partial_t\underline{\partial}_b v + \underline{A}^{a0c}\underline{\partial}_c u\underline{\partial}_a\partial_t v + \underline{A}^{abc}\underline{\partial}_a u\underline{\partial}_b\partial_c v \\
=:& R_3(u,v)+R_4(u,v),
\end{aligned}$$

where

$$R_3(u,v) := \underline{A}^{000}\partial_t u \partial_t \partial_t v$$

and $R_4(u,v)$ denotes the remaining terms. We have $Z^I R_1 = Z^I R_3 + Z^I R_4$. To estimate $|Z^I R_3|$, we observe that $A^{\alpha\beta\gamma}$ is a null cubic form so by Proposition 4.1.1, $|Z^I A^{000}| \leqslant C(s/t)^2$. We find

$$Z^I\big(\underline{A}^{000}\partial_t u \partial_t \partial_t v\big) = \sum_{I_1+I_2+I_3=I} Z^{I_3} A^{000}\, Z^{I_1}\partial_t u\, Z^{I_2}\partial_t\partial_t v,$$

so that

$$\big|Z^I\big(\underline{A}^{000}\partial_t u \partial_t \partial_t v\big)\big| \leqslant C(n,|I|)K(s/t)^2 \sum_{|I_1|+|I_2|\leqslant|I|} |Z^{I_1}\partial_t u\, Z^{I_2}\partial_t\partial_t v|.$$

To see the estimates on $Z^I R_4$ terms, we just remark that by Lemma 2.2.2, $|Z^I \underline{A}^{\alpha\beta\gamma}| \leqslant C(n,I)K$. We can control $Z^I R_3$ by Ω_1.

Finally, by combining the estimate on R_2, R_3 and R_4, the desired result is established. □

The fourth type of terms we need to control is the commutator between Z^I and a null quadratic form.

Proposition 4.3.2. *Consider a null quadratic form acting on the gradient of function u and on the Hessian matrix of v, that is, $A^{\alpha\beta\gamma}\partial_\gamma u \partial_\alpha \partial_\beta v$, for functions u,v defined in the cone $\mathcal{K}$. The following estimate holds:*

$$\begin{aligned}
\big|[Z^I, A^{\alpha\beta\gamma}\partial_\gamma u \partial_\alpha\partial_\beta]v\big| \leqslant& CK(s/t)^2 \sum_{\substack{|I_2|+|I_3|\leqslant|I|,\\ |I_3|<|I|}} |Z^{I_2}\partial_t u Z^{I_3}\partial_t\partial_t v| \\
&+ CK\sum_{c,\alpha,\beta}\sum_{\substack{|I_2|+|I_3|\leqslant|I|,\\ |I_3|<|I|}} |Z^{I_2}\underline{\partial}_c u Z^{I_3}\underline{\partial}_\alpha\underline{\partial}_\beta v| \\
&+ CK\sum_{c,\alpha,\beta}\sum_{\substack{|I_2|+|I_3|\leqslant|I|,\\ |I_3|<|I|}} |Z^{I_2}\underline{\partial}_\alpha u Z^{I_3}\underline{\partial}_c\underline{\partial}_\beta v| \\
&+ CK\sum_{c,\alpha,\beta}\sum_{\substack{|I_2|+|I_3|\leqslant|I|,\\ |I_3|<|I|}} |Z^{I_2}\underline{\partial}_\alpha u Z^{I_3}\underline{\partial}_\beta\underline{\partial}_c v| \\
&+ t^{-1}CK\sum_{\alpha,\gamma}\sum_{\substack{|I_2|\leqslant|I|-1\\ |I_1|+|I_2|\leqslant|I|}} |\partial_\gamma Z^{I_1}u \partial_\alpha Z^{I_2}v|.
\end{aligned}$$

Proof. We use the change of frame formula

$$A^{\alpha\beta\gamma}\partial_\gamma u\partial_\alpha\partial_\beta v = \underline{A}^{\alpha\beta\gamma}\underline{\partial}_\gamma u\,\underline{\partial}_\alpha\underline{\partial}_\beta v + A^{\alpha\beta\gamma}\partial_\gamma u\,\partial_\alpha\big(\Psi_\beta^{\beta'}\big)\underline{\partial}_{\beta'} v$$

and find

$$\begin{aligned}&[Z^I, A^{\alpha\beta\gamma}\partial_\gamma u\partial_\alpha\partial_\beta]v\\&= [Z^I, \underline{A}^{\alpha\beta\gamma}\underline{\partial}_\gamma u\underline{\partial}_\alpha\underline{\partial}_\beta]v + [Z^I, A^{\alpha\beta\gamma}\partial_\gamma u\partial_\alpha(\Psi_\beta^{\beta'})\underline{\partial}_{\beta'}]v\\&=: R_1(I,u,v) + R_2(I,u,v).\end{aligned}$$

We first decompose $R_1(I,u,v)$ as

$$\begin{aligned}&[Z^I, \underline{A}^{\alpha\beta\gamma}\underline{\partial}_\gamma u\underline{\partial}_\alpha\underline{\partial}_\beta]v\\&= \sum_{\substack{I_2+I_3+I_4=I,\\|I_3|<|I|}} Z^{I_4}\underline{A}^{\alpha\beta\gamma}Z^{I_2}(\underline{\partial}_\gamma u)Z^{I_3}(\underline{\partial}_\alpha\underline{\partial}_\beta v) + \underline{A}^{\alpha\beta\gamma}\underline{\partial}_\gamma u[Z^I, \underline{\partial}_\alpha\underline{\partial}_\beta]v\\&=: R_3(I,u,v) + R_4(I,u,v),\end{aligned}$$

while $R_3(I,u,v)$ is decomposed as

$$\begin{aligned}&R_3(I,u,v)\\&= \sum_{\substack{I_2+I_3+I_4=I,\\|I_3|<|I|}} Z^{I_4}\underline{A}^{000}Z^{I_2}\partial_t u Z^{I_3}\partial_t\partial_t v\\&\quad+ \sum_{\substack{I_2+I_3+I_4=I,\\|I_3|<|I|}} Z^{I_4}\underline{A}^{00c}Z^{I_2}\underline{\partial}_c u Z^{I_3}\partial_t\partial_t v + \sum_{\substack{I_2+I_3+I_4=I,\\|I_3|<|I|}} Z^{I_4}\underline{A}^{0b0}Z^{I_2}\partial_t u Z^{I_3}\partial_t\underline{\partial}_c v\\&\quad+ \sum_{\substack{I_2+I_3+I_4=I,\\|I_3|<|I|}} Z^{I_4}\underline{A}^{a00}Z^{I_2}\partial_t u Z^{I_3}\underline{\partial}_a\partial_t v + \sum_{\substack{I_2+I_3+I_4=I,\\|I_3|<|I|}} Z^{I_4}\underline{A}^{0bc}Z^{I_2}\underline{\partial}_c u Z^{I_3}\partial_t\underline{\partial}_b v\\&\quad+ \sum_{\substack{I_2+I_3+I_4=I,\\|I_3|<|I|}} Z^{I_4}\underline{A}^{a0c}Z^{I_2}\underline{\partial}_c u Z^{I_3}\underline{\partial}_a\partial_t v + \sum_{\substack{I_2+I_3+I_4=I,\\|I_3|<|I|}} Z^{I_4}\underline{A}^{ab0}Z^{I_2}\partial_t u Z^{I_3}\underline{\partial}_a\underline{\partial}_b v\\&\quad+ \sum_{\substack{I_2+I_3+I_4=I,\\|I_3|<|I|}} Z^{I_4}\underline{A}^{abc}Z^{I_2}\underline{\partial}_c u Z^{I_3}\underline{\partial}_a\underline{\partial}_b v.\end{aligned}$$

We observe that $\mathcal{A}$ is a null cubic form and, by Proposition 4.1.1, and therefore

$$|Z^{I_4}\underline{A}^{000}| \leqslant C(|I|)K(s/t)^2.$$

By Lemma 2.2.2, we have

$$|Z^{I_4}\underline{A}^{\alpha\beta\gamma}| \leqslant K.$$

The term $R_3(I,u,v)$ is estimated as follows:

$$
\begin{aligned}
&\big|R_3(I,u,v)\big| \\
&\leqslant CK(s/t)^2 \sum_{\substack{|I_2|+|I_3|\leqslant|I|,\\ |I_3|<|I|}} \big|Z^{I_2}\partial_t u Z^{I_3}\partial_t\partial_t v\big| \\
&\quad + CK \sum_{\substack{|I_2|+|I_3|\leqslant|I|,\\ |I_3|<|I|}} \big|Z^{I_2}\underline{\partial}_c u Z^{I_3}\partial_t\partial_t v\big| + CK \sum_{\substack{|I_2|+|I_3|\leqslant|I|,\\ |I_3|<|I|}} \big|Z^{I_2}\partial_t u Z^{I_3}\partial_t\underline{\partial}_c v\big| \\
&\quad + CK \sum_{\substack{|I_2|+|I_3|\leqslant|I|,\\ |I_3|<|I|}} \big|Z^{I_2}\partial_t u Z^{I_3}\underline{\partial}_a\partial_t v\big| + CK \sum_{\substack{|I_2|+|I_3|\leqslant|I|,\\ |I_3|<|I|}} \big|Z^{I_2}\underline{\partial}_c u Z^{I_3}\partial_t\underline{\partial}_b v\big| \\
&\quad + CK \sum_{\substack{|I_2|+|I_3|\leqslant|I|,\\ |I_3|<|I|}} \big|Z^{I_2}\underline{\partial}_c u Z^{I_3}\underline{\partial}_a\partial_t v\big| + CK \sum_{\substack{|I_2|+|I_3|\leqslant|I|,\\ |I_3|<|I|}} \big|Z^{I_2}\partial_t u Z^{I_3}\underline{\partial}_a\underline{\partial}_b v\big| \\
&\quad + CK \sum_{\substack{|I_2|+|I_3|\leqslant|I|,\\ |I_3|<|I|}} \big|Z^{I_2}\underline{\partial}_c u Z^{I_3}\underline{\partial}_a\underline{\partial}_b v\big|,
\end{aligned}
$$

so that

$$
\begin{aligned}
\big|R_3(I,u,v)\big| \leqslant& CK(s/t)^2 \sum_{\substack{|I_2|+|I_3|\leqslant|I|,\\ |I_3|<|I|}} \big|Z^{I_2}\partial_t u Z^{I_3}\partial_t\partial_t v\big| \\
&+ CK \sum_{c,\alpha,\beta} \sum_{\substack{|I_2|+|I_3|\leqslant|I|,\\ |I_3|<|I|}} \big|Z^{I_2}\underline{\partial}_c u Z^{I_3}\underline{\partial}_\alpha\underline{\partial}_\beta v\big| \\
&+ CK \sum_{c,\alpha,\beta} \sum_{\substack{|I_2|+|I_3|\leqslant|I|,\\ |I_3|<|I|}} \big|Z^{I_2}\underline{\partial}_\alpha u Z^{I_3}\underline{\partial}_c\underline{\partial}_\beta v\big| \\
&+ CK \sum_{c,\alpha,\beta} \sum_{\substack{|I_2|+|I_3|\leqslant|I|,\\ |I_3|<|I|}} \big|Z^{I_2}\underline{\partial}_\alpha u Z^{I_3}\underline{\partial}_\beta\underline{\partial}_c v\big|.
\end{aligned}
$$

The term $R_4(I,u,v)$ is also decomposed as:

$$
\begin{aligned}
R_4(I,u,v) =& \underline{A}^{000}\partial_t u[Z^I,\partial_t\partial_t]v + \underline{A}^{00c}\underline{\partial}_c u[Z^I,\partial_t\partial_t]v \\
&+ \underline{A}^{0b0}\partial_t u[Z^I,\partial_t\underline{\partial}_b]v + \underline{A}^{a00}\partial_t u[Z^I,\underline{\partial}_a\partial_t]v \\
&+ \underline{A}^{0bc}\underline{\partial}_c u[Z^I,\partial_t\underline{\partial}_b]v + \underline{A}^{a0c}\underline{\partial}_c u[Z^I,\underline{\partial}_a\partial_t]v \\
&+ \underline{A}^{ab0}\partial_t u[Z^I,\underline{\partial}_a\underline{\partial}_b]v + \underline{A}^{abc}\underline{\partial}_c u[Z^I,\underline{\partial}_a\underline{\partial}_b]v.
\end{aligned}
$$

Thanks to the null condition, we have

$$
\begin{aligned}
\big|R_4(I,u,v)\big| \leqslant& CK(s/t)^2\partial_t u\big|[Z^I,\partial_t\partial_t]v\big| + CK\big|\underline{\partial}_c u[Z^I,\partial_t\partial_t]v\big| \\
&+ CK\big|\partial_t u[Z^I,\partial_t\underline{\partial}_b]v\big| + CK\big|\partial_t u[Z^I,\underline{\partial}_a\partial_t]v\big| \\
&+ CK\big|\underline{\partial}_c u[Z^I,\partial_t\underline{\partial}_b]v\big| + CK\big|\underline{\partial}_c u[Z^I,\underline{\partial}_a\partial_t]v\big| \\
&+ CK\big|\partial_t u[Z^I,\underline{\partial}_a\underline{\partial}_b]v\big| + CK\big|\underline{\partial}_c u[Z^I,\underline{\partial}_a\underline{\partial}_b]v\big|.
\end{aligned}
$$

Observe now that thanks to the commutator estimates (3.3.3) and (3.3.4), we have

$$\big|\underline{\partial}_\alpha u[Z^I,\partial_t\partial_t]v\big| \leqslant C \sum_{\substack{\alpha,\beta\\ |I'|\leqslant|I|-1}} \big|\underline{\partial}_\alpha u\partial_\alpha\partial_\beta Z^{I'}v\big|,$$

$$\big|\underline{\partial}_\alpha u[Z^I,\underline{\partial}_a\underline{\partial}_\beta]v\big| \leqslant C \sum_{\substack{a,\beta\\ |I'|\leqslant|I|-1}} \big|\underline{\partial}_\alpha u\underline{\partial}_a\underline{\partial}_\beta Z^{I'}v\big| + Ct^{-1}\sum_{\substack{\gamma\\ |I'|<|I|}} \big|\underline{\partial}_\alpha u\partial_\gamma Z^{I'}v\big|$$

and

$$\big|\underline{\partial}_\alpha u[Z^I,\underline{\partial}_\beta\underline{\partial}_c]v\big| \leqslant C \sum_{\substack{\beta,c\\ |I'|\leqslant|I|-1}} \big|\underline{\partial}_\alpha u\underline{\partial}_c\underline{\partial}_\beta Z^{I'}v\big| + Ct^{-1}\sum_{\substack{\gamma\\ |I'|<|I|}} \big|\underline{\partial}_\alpha u\partial_\gamma Z^{I'}v\big|.$$

We also note that $\sum_\alpha \big|\underline{\partial}_\alpha u\big| \leqslant C\sum_\alpha \big|\partial_\alpha u\big|$, so that $\big|R_4(I,u,v)\big|$ is bounded by

$$\begin{aligned}
&\big|R_4(I,u,v)\big|\\
&\leqslant CK(s/t)^2 \sum_{\substack{\alpha,\beta\\ |I'|\leqslant|I|-1}} \big|\partial_t u\partial_\alpha\partial_\beta Z^{I'}v\big| + CK\sum_{\substack{\alpha,\beta,c\\ |I'|\leqslant|I|-1}} \big|\underline{\partial}_c u\partial_\alpha\partial_\beta Z^{I'}v\big|\\
&\quad + CK\sum_{\substack{\alpha,\beta,c\\ |I'|\leqslant|I|-1}} \big|\partial_\alpha u\underline{\partial}_c\underline{\partial}_\beta v\big| + CKt^{-1}\sum_{\substack{\alpha,\gamma\\ |I'|\leqslant|I|-1}} \big|\partial_\gamma u\partial_\alpha Z^{I'}v\big|.
\end{aligned}$$

In conclusion, $R_1(I,u,v)$ is bounded by

$$\begin{aligned}
\big|R_1(I,u,v)\big| \leqslant& CK(s/t)^2 \sum_{\substack{|I_2|+|I_3|\leqslant|I|,\\ |I_3|<|I|}} \big|Z^{I_2}\partial_t u Z^{I_3}\partial_t\partial_t v\big|\\
&+ CK\sum_{c,\alpha,\beta}\sum_{\substack{|I_2|+|I_3|\leqslant|I|,\\ |I_3|<|I|}} \big|Z^{I_2}\underline{\partial}_c u Z^{I_3}\underline{\partial}_\alpha\underline{\partial}_\beta v\big|\\
&+ CK\sum_{c,\alpha,\beta}\sum_{\substack{|I_2|+|I_3|\leqslant|I|,\\ |I_3|<|I|}} \big|Z^{I_2}\underline{\partial}_\alpha u Z^{I_3}\underline{\partial}_c\underline{\partial}_\beta v\big|\\
&+ CK\sum_{c,\alpha,\beta}\sum_{\substack{|I_2|+|I_3|\leqslant|I|,\\ |I_3|<|I|}} \big|Z^{I_2}\underline{\partial}_\alpha u Z^{I_3}\underline{\partial}_\beta\underline{\partial}_c v\big|\\
&+ t^{-1}CK\sum_{\substack{\alpha,\gamma\\ |I'|\leqslant|I|-1}} \big|\partial_\gamma u\partial_\alpha Z^{I'}v\big|.
\end{aligned}$$

Next, we turn our attention to the estimate of $R_2(I,u,v)$:

$$\begin{aligned}
R_2(I,u,v) =& A^{\alpha\beta\gamma}\partial_\gamma u\,\partial_\alpha\Psi_\beta^{\beta'}[Z^I,\underline{\partial}_{\beta'}]v\\
&+ \sum_{\substack{I_1+I_2+I_3=I,\\ |I_2|<|I|}} A^{\alpha\beta\gamma}Z^{I_1}\partial_\gamma u\,Z^{I_2}\underline{\partial}_{\beta'}v\,Z^{I_3}\partial_\alpha\Psi_\beta^{\beta'}\\
=:& R_5(I,u,v) + R_6(I,u,v).
\end{aligned}$$

To estimate the term $R_5(I,u,v)$, we recall (3.2.1) and the fact that $|\partial_\alpha \Psi_\beta^{\beta'}| \leqslant Ct^{-1}$, so that

$$\big|A^{\alpha\beta\gamma}\partial_\gamma u\,\partial_\alpha\Psi_\beta^{\beta'}[Z^I,\underline{\partial}_{\beta'}]v\big| \leqslant CKt^{-1}\sum_{\substack{\beta,\gamma\\|I'|\leqslant|I|-1}}\big|\partial_\gamma u\partial_\beta Z^{I'}v\big|.$$

In the same way, for $R_6(I,u,v)$ we have

$$R_6(I,u,v) \leqslant CKt^{-1}\sum_{\beta,\gamma}\sum_{\substack{|I_1|+|I_2|\leqslant|I|\\|I_2|<|I|}}\big|\partial_\beta Z^{I_1}u\partial_\gamma Z^{I_2}v\big|.$$

This establishes the desired result. □

Chapter 5

Sobolev and Hardy inequalities on hyperboloids

5.1 A Sobolev inequality on hyperboloids

To turn L^2 energy estimates into L^∞ estimates, we will rely on the following Sobolev inequality.

Proposition 5.1.1 (Sobolev-type estimate on hyperboloids). *Let u be a sufficiently regular function defined in the cone $\mathcal{K} = \{|x| < t-1\}$, then for all $s > 0$ (and with $t = \sqrt{s^2 + |x|^2}$)*

$$\sup_{\mathcal{H}_s} t^{3/2}|u(t,x)| \leqslant C \sum_L \sum_{|I|\leqslant 2} \|L^I u\|_{L^2(\mathcal{H}_s)}, \tag{5.1.1}$$

where $C > 0$ is a universal constant and the summation in L is over all vector fields $L_a = x^a\partial_t + t\partial_a$, $a = 1,2,3$.

In comparison to Lemma 7.6.1 in Hörmander (1997), observe that the right-hand side of (5.1.1) does not contain the rotation fields $\Omega_{ab} := x^a\partial_b - x^b\partial_a$.

Proof. Recall the relation $t = \sqrt{s^2 + |x|^2}$ on $\mathcal{H}_s$ and consider a function u defined in $\mathcal{K}$, its restriction to the hyperboloid $\mathcal{H}_s$ is, by definition,

$$w_s(x) := u(\sqrt{s^2 + |x|^2}, x).$$

Fix s_0 and a point (t_0, x_0) on the hyperboloid $\mathcal{H}_{s_0}$, with $t_0 = \sqrt{s_0^2 + |x_0|^2}$. Observe that

$$\partial_a w_{s_0}(x) = \underline{\partial}_a u\big(\sqrt{s_0^2 + |x|^2}, x\big) = \underline{\partial}_a u(t,x), \tag{5.1.2}$$

with $t = \sqrt{s_0^2 + |x|^2}$ and, therefore,

$$t\partial_a w_{s_0}(x) = t\underline{\partial}_a u\big(\sqrt{s_0^2 + |x|^2}, t\big) = L_a u(t,x). \tag{5.1.3}$$

Introduce the function $g_{s_0,t_0}(y) := w_{s_0}(x_0 + t_0\, y)$ and note that

$$g_{s_0,t_0}(0) = w_{s_0}(x_0) = u\big(\sqrt{s_0^2 + |x_0|^2}, x_0\big) = u(t_0, x_0).$$

Applying the standard Sobolev inequality to g_{s_0,t_0}, we obtain

$$\big|g_{s_0,t_0}(0)\big|^2 \leqslant C \sum_{|I|\leqslant 2} \int_{B(0,1/3)} |\partial^I g_{s_0,t_0}(y)|^2\, dy,$$

where $B(0,1/3) \subset \mathbb{R}^3$ denotes the ball centered at the origin and with radius $1/3$.

Taking into account the identity (with $x = x_0 + t_0 y$)

$$\begin{aligned}\partial_a g_{s_0,t_0}(y) &= t_0 \partial_a w_{s_0}(x_0 + t_0 y) = t_0 \partial_a w_{s_0}(x)\\ &= t_0 \underline{\partial}_a u(t,x)\big),\end{aligned}$$

in view of (5.1.2), we see that for all I

$$\partial^I g_{s_0,t_0}(y) = (t_0 \underline{\partial})^I u(t,x)$$

and, therefore,

$$\begin{aligned}\big|g_{s_0,t_0}(0)\big|^2 \leqslant& C \sum_{|I|\leqslant 2} \int_{B(0,1/3)} \big|(t_0\underline{\partial})^I u(t,x)\big)\big|^2 dy\\ =& C t_0^{-3} \sum_{|I|\leqslant 2} \int_{B((t_0,x_0),t_0/3)\cap \mathcal{H}_{s_0}} \big|(t_0\underline{\partial})^I u(t,x)\big)\big|^2 dx.\end{aligned}$$

We can check that

$$\begin{aligned}(t_0\underline{\partial}_a(t_0\underline{\partial}_b w_{s_0})) &= t_0^2 \underline{\partial}_a \underline{\partial}_b w_{s_0}\\ &= (t_0/t)^2 (t\underline{\partial}_a)(t\underline{\partial}_b) w_{s_0} - (t_0/t)^2 (x^a/t) L_b w_{s_0}.\end{aligned}$$

We also remark that $x^a/t = x_0^a/t + y t_0/t = (x_0^a/t_0 + y)(t_0/t)$, so that, in the region $y \in B(0,1/3)$ of interest, the factor $|x^a/t|$ is bounded by $C(t_0/t)$. We conclude that for any $|I| \leqslant 2$,

$$|(t_0\underline{\partial})^I u| \leqslant \sum_{|J|\leqslant |I|} |L^I u| (t_0/t)^I.$$

On the other hand, when $|x_0| \leqslant t_0/2$ then $t_0 \leqslant \frac{2}{\sqrt{3}} s_0$ and thus

$$t_0 \leqslant C s_0 \leqslant C\sqrt{|x|^2 + s_0^2} = Ct,$$

C being a universal constant thoughout. When $|x_0| \geqslant t_0/2$ then in the region $B((t_0,x_0),t_0/3) \cap \mathcal{H}_{s_0}$ we have also

$$t_0 \leqslant C|x| \leqslant C\sqrt{|x|^2 + s_0^2} = Ct.$$

Consequently, it follows that

$$|(t_0\underline{\partial})^I u| \leqslant C \sum_{|J|\leqslant|I|} |L^I u| \tag{5.1.4}$$

and

$$\begin{aligned}|g_{s_0,t_0}(y_0)|^2 \leqslant& Ct_0^{-3} \sum_{|I|\leqslant 2} \int_{B(x_0,t_0/3)\cap\mathcal{H}_{s_0}} \left|(t\underline{\partial})^I u(t,x))\right|^2 dx\\ \leqslant& Ct_0^{-3} \sum_{|I|\leqslant 2} \int_{\mathcal{H}_{s_0}} \left|L^I u(t,x)\right|^2 dx,\end{aligned}$$

which completes the proof of Proposition 5.1.1. □

5.2 Application of the Sobolev inequality on hyperboloids

Using now the hyperboloidal energy defined in (2.3.2) and combining Proposition 5.1.1 with the technical estimate (3.3.6), we can deduce various sup-norm estimates, presented now. For clarity in the presention, we make use of the notation $E_{m,\sigma}$ (defined in Chapter 2) in order to emphasize the dependency of the hyperboloidal energy upon the coeffficient σ in the Klein-Gordon equation.

Lemma 5.2.1 (L^∞ estimates on derivatives up to first-order). *If u is a sufficiently regular function supported in $\mathcal{K}$, then the following estimates hold:*

$$\begin{aligned}&(a) \quad \sup_{\mathcal{H}_s} \left|t^{1/2} s\partial_\alpha u\right| \leqslant C \sum_{|I|\leqslant 2} E_m(s, Z^I u)^{1/2},\\ &(b) \quad \sup_{\mathcal{H}_s} \left|t^{3/2}\underline{\partial}_a u\right| \leqslant C \sum_{|I|\leqslant 2} E_m(s, Z^I u)^{1/2},\\ &(c) \quad \sup_{\mathcal{H}_s} \left|\sigma\, t^{3/2} u\right| \leqslant C \sum_{|I|\leqslant 2} E_{m,\sigma}(s, Z^I u)^{1/2},\end{aligned} \tag{5.2.1}$$

where Z stands for any admissible vector field, that is, any of ∂_α, L_a, and C is a universal constant.

Remark 5.2.1. Let us illustrate our result with the homogeneous linear wave equation

$$\Box w = 0, \quad w|_{\mathcal{H}_{B+1}} = w_0, \quad \partial_t w|_{\mathcal{H}_{B+1}} = w_1, \tag{5.2.2}$$

where the solution w_i is defined in $\mathcal{H}_{B+1} \cap \mathcal{K}$. Thanks to the energy estimate in Proposition 2.3.1, the energy $E_m(s, Z^I w)$ is controlled by the initial energy. By the commutator estimates (3.3.1) and (3.3.2) and by the Sobolev inequality, we find

$$|\underline{\partial}_a w| \leqslant Ct^{-3/2}, \quad |\partial_\alpha w| \leqslant Ct^{-3/2+1}(t^2 - r^2)^{-1/2}, \tag{5.2.3}$$

which are classical estimates. We emphasize that our argument of proof is "robust" in the sense that it uses neither the explicit expression of the solution nor the scaling vector field $S = r\partial_r + t\partial_t$.

Now, we turn our attention to the energy and decay estimates for the "good" second-order derivatives, that is, derivatives such as $\underline{\partial}_a \partial_\alpha u$.

Lemma 5.2.2 (Bounds on second-order derivatives). *For every sufficiently regular function u supported in the cone $\mathcal{K}$, the following estimates hold:*

$$\sup_{\mathcal{H}_s} |t^{3/2} s \underline{\partial}_a \underline{\partial}_\alpha u| + \sup_{\mathcal{H}_s} |t^{3/2} s \underline{\partial}_\alpha \underline{\partial}_a u| \leqslant C \sum_{|I| \leqslant 3} E_m(s, Z^I u)^{1/2}, \tag{5.2.4}$$

$$\int_{\mathcal{H}_s} |s \underline{\partial}_a \underline{\partial}_\alpha u|^2 dx + \int_{\mathcal{H}_s} |s \underline{\partial}_\alpha \underline{\partial}_a u|^2 dx \leqslant C \sum_{|I| \leqslant 1} E_m(s, Z^I u). \tag{5.2.5}$$

Proof. Recalling that $\underline{\partial}_a = t^{-1} L_a$, we obtain $|\underline{\partial}_a \partial_\alpha u| \leqslant t^{-1} |L_a \partial u|$. By Lemma 5.2.1 and the commutator estimate (3.1.5), we obtain (5.2.4). The second estimate is immediate in view of the expression (2.3.2). □

Remark 5.2.2. Energy estimates and L^∞ estimates for the second-order time derivative $\underline{\partial}_0 \underline{\partial}_0 u$ will be derived later from the wave equation itself, thanks to the decomposition in Proposition 2.2.1.

At the end of this section, we state the L^∞ estimates of the solution of wave equation (i.e. $c_i = 0$).

Lemma 5.2.3. *If u is a sufficiently regular function supported in the cone $\mathcal{K}$, then for any multi-index J, if $\sum_{|I| \leqslant |J|+2} E_m(s, Z^I u)^{1/2} \leqslant C' s^\delta$ for some $\delta \geqslant 0$, then one has*

$$|Z^J u| \leqslant CC' t^{(-2+\delta)/2} (t-r)^{(1+\delta)/2} \leqslant C'' C' t^{-3/2} s^{1+\delta}, \tag{5.2.6}$$

where C, C', C'' are universal constants.

Proof. Using that $\sum_{|I| \leqslant |J|+2} E_m(s, Z^I u)^{1/2}$ is bounded by $C' s^\delta$ and recalling by Lemma 5.2.1, we find in the cone $\mathcal{K}$

$$|\partial_r Z^I u| \leqslant Ct^{-(2-\delta)/2} (t-r)^{-(1-\delta)/2}.$$

Then, (5.2.6) follows by integration along radial directions. □

5.3 Hardy inequality for the hyperboloidal foliation

In this section we establish an analogue of the classical Hardy inequality but generalized to hyperboloidal foliations. This inequality will be used in order to control L^2 norms of wave components such as $\|Z^I u\|_{L^2(\mathcal{H}_s)}$.

Proposition 5.3.1 (The hyperboloidal Hardy inequality). *For all sufficiently regular functions u supported in the cone $\mathcal{K}$, one has*

$$\begin{aligned}\|s^{-1}u\|_{L^2(\mathcal{H}_s)} \leqslant& C\|s_0^{-1}u\|_{L^2(\mathcal{H}_{s_0})} + C\sum_a \|\underline{\partial}_a u\|_{L^2(\mathcal{H}_s)} \\ &+ C\sum_a \int_{s_0}^{s} \tau^{-1}\Big(\|(\tau/t)\partial_a u\|_{L^2(\mathcal{H}_\tau)} + \|\underline{\partial}_a u\|_{L^2(\mathcal{H}_\tau)}\Big)\,d\tau.\end{aligned} \tag{5.3.1}$$

Before proving this result, we begin with the following modified version of the classical Hardy inequality.

Lemma 5.3.1. *For all sufficiently regular functions u supported in the cone $\mathcal{K}$, one has*

$$\|r^{-1}u\|_{L^2(\mathcal{H}_s)} \leqslant C\sum_a \|\underline{\partial}_a u\|_{L^2(\mathcal{H}_s)}.$$

Proof. As in the proof of Proposition 5.1.1, we consider the function $w_s(x) := u\big(\sqrt{s^2+|x|^2}, x\big)$, which satisfies

$$\partial_a w_s(x) = \underline{\partial}_a u\big(\sqrt{s^2+|x|^2}, x\big).$$

We then apply the classical Hardy inequality to w_s and obtain

$$\begin{aligned}\int_{\mathbb{R}^3} |r^{-1}w_s(x)|^2 dx \leqslant C\int_{\mathbb{R}^3} |\nabla w_s(x)|^2 dx &= C\sum_a \int_{\mathbb{R}^3} \big|\underline{\partial}_a u(\sqrt{s^2+r^2}, x)\big|^2 dx \\ &\leqslant C\sum_a \int_{\mathcal{H}_s} \big|\underline{\partial}_a u(t,x)\big|^2 dx.\end{aligned}$$

□

Proof of Proposition 5.3.1. We introduce a smooth cut-off function χ satisfying

$$\chi(r) = \begin{cases} 1, & 2/3 \leqslant r, \\ 0, & 0 \leqslant r \leqslant 1/3 \end{cases}$$

and consider the decomposition

$$\|s^{-1}u\|_{L^2(\mathcal{H}_s)} \leqslant \|\chi(r/t)s^{-1}u\|_{L^2(\mathcal{H}_s)} + \|(1-\chi(r/t))s^{-1}u\|_{L^2(\mathcal{H}_s)}$$

which distinguish between the region "near" and "away" from the light cone.

The estimate of $\|(1-\chi(r/t))s^{-1}u\|_{L^2(\mathcal{H}_s)}$ is based on the following observation:

$$(1-\chi(r/t))s^{-1} \leqslant Ct^{-1} \qquad \text{in the cone } \mathcal{K},$$

so that, by Lemma 5.3.1,

$$\begin{aligned}\|(1-\chi(r/t))us^{-1}\|_{L^2(\mathcal{H}_s)} &\leqslant \|t^{-1}u\|_{L^2(\mathcal{H}_s)} \\ &\leqslant \|r^{-1}u\|_{L^2(\mathcal{H}_s)} \leqslant C\sum_a \|\underline{\partial}_a u\|_{L^2(\mathcal{H}_s)}.\end{aligned} \tag{5.3.2}$$

The estimate near the light cone is more delicate and to deal with the term $\|\chi(r/t)s^{-1}u\|_{L^2(\mathcal{H}_s)}$, we proceed as follows: in the region $\mathcal{K}_{[s_0,s]}$, we can find a positive constant C

$$\chi(r/t) \leqslant C\frac{\chi(r/t)r}{(1+r^2)^{1/2}},$$

and thus

$$\|\chi(r/t)s^{-1}u\|_{L^2(\mathcal{H}_s)} \leqslant C\|r(1+r^2)^{-1/2}\chi(r/t)s^{-1}u\|_{L^2(\mathcal{H}_s)}.$$

So, we can focus on controlling this latter term.

To this end, we consider the vector field

$$W = \left(0, -x^a\frac{t(u\chi(r/t))^2}{(1+r^2)s^2}\right)$$

defined in $\mathcal{K}$ and we compute its divergence

$$\begin{aligned}&\operatorname{div} W \\ &= s^{-1}\partial_a u\frac{r\chi(r/t)u}{(1+r^2)^{1/2}s}\frac{-2x^a t\chi(r/t)}{r(1+r^2)^{1/2}} - s^{-1}\frac{u}{r}\frac{r\chi(r/t)u}{s(1+r^2)^{1/2}}\frac{2\chi'(r/t)r}{(1+r^2)^{1/2}} \\ &\quad - \left(\frac{r^2t+3t}{(1+r^2)^2s^2} + \frac{2r^2t}{(1+r^2)s^4}\right)\left(u\chi(r/t)\right)^2.\end{aligned}$$

Next, we integrate the above inequality in the region $\mathcal{K}_{[s_0,s_1]} \subset \mathcal{K} \cap \{s_0 \leqslant \sqrt{t^2-r^2} \leqslant s_1\}$ with respect to the Lebesgue measure in $\mathbb{R}^4$:

$$\begin{aligned}&\int_{\mathcal{K}_{[s_0,s_1]}} \operatorname{div} W dxdt \\ &= -2\int_{\mathcal{K}_{[s_0,s_1]}} s^{-1}\left(\partial_a u\frac{r\chi(r/t)u}{(1+r^2)^{1/2}s}\frac{x^a t\chi(r/t)}{r(1+r^2)^{1/2}}\right)dxdt \\ &\quad - 2\int_{\mathcal{K}_{[s_0,s_1]}} s^{-1}\frac{u}{r}\frac{r\chi(r/t)u}{s(1+r^2)^{1/2}}\frac{\chi'(r/t)r}{(1+r^2)^{1/2}}dxdt \\ &\quad - \int_{\mathcal{K}_{[s_0,s_1]}}\left(\frac{r^2t+3t}{(1+r^2)^2s^2} + \frac{2r^2t}{(1+r^2)s^4}\right)\left(u\chi(r/t)\right)^2 dxdt\end{aligned}$$

thus

$$\begin{aligned}
&\int_{\mathcal{K}_{[s_0,s_1]}} \operatorname{div} W dxdt\\
&= -2\int_{s_0}^{s_1}\int_{\mathcal{H}_s}(s/t)s^{-1}\Big(\partial_a u\frac{r\chi(r/t)u}{(1+r^2)^{1/2}s}\frac{x^a t\chi(r/t)}{r(1+r^2)^{1/2}}\Big)dxds\\
&\quad -2\int_{s_0}^{s_1}\int_{\mathcal{H}_s}(s/t)s^{-1}\frac{u}{r}\frac{r\chi(r/t)u}{s(1+r^2)^{1/2}}\frac{\chi'(r/t)r}{(1+r^2)^{1/2}}dxds\\
&\quad -\int_{s_0}^{s_1}\int_{\mathcal{H}_s}(s/t)\Big(\frac{r^2t+3t}{(1+r^2)^2s^2}+\frac{2r^2t}{(1+r^2)s^4}\Big)\big(u\chi(r/t)\big)^2dxds\\
&=: \int_{s_0}^{s_1}(T_1+T_2+T_3)\,ds,
\end{aligned}$$

where

$$\begin{aligned}
T_1 =& -2s^{-1}\int_{\mathcal{H}_s}(s/t)\Big(\partial_a u\frac{r\chi(r/t)u}{(1+r^2)^{1/2}s}\frac{x^a t\chi(r/t)}{r(1+r^2)^{1/2}}\Big)dx\\
\leqslant& 2s^{-1}\Big\|\frac{ru\chi(r/t)}{s(1+r^2)^{1/2}}\Big\|_{L^2(\mathcal{H}_s)}\\
&\qquad\cdot\sum_a\|(s/t)\partial_a u\|_{L^2(\mathcal{H}_s)}\big\|\chi(r/t)x^a tr^{-1}(1+r^2)^{-1/2}\big\|_{L^\infty(\mathcal{H}_s)}\\
\leqslant& Cs^{-1}\Big\|\frac{ru\chi(r/t)}{s(1+r^2)^{1/2}}\Big\|_{L^2(\mathcal{H}_s)}\sum_a\|(s/t)\partial_a u\|_{L^2(\mathcal{H}_s)},
\end{aligned}$$

$$\begin{aligned}
T_2 =& -2s^{-1}\int_{\mathcal{H}_s}(s/t)\frac{u}{r}\frac{r\chi(r/t)u}{s(1+r^2)^{1/2}}\frac{\chi'(r/t)r}{(1+r^2)^{1/2}}dx\\
\leqslant& Cs^{-1}\Big\|\frac{ru\chi(r/t)}{s(1+r^2)^{1/2}}\Big\|_{L^2(\mathcal{H}_s)}\|ur^{-1}\|_{L^2(\mathcal{H}_s)}\big\|r\chi'(r/t)(1+r^2)^{-1/2}\big\|_{L^\infty(\mathcal{H}_s)}\\
\leqslant& Cs^{-1}\Big\|\frac{ru\chi(r/t)}{s(1+r^2)^{1/2}}\Big\|_{L^2(\mathcal{H}_s)}\sum_a\|\underline{\partial}_a u\|_{L^2(\mathcal{H}_s)},
\end{aligned}$$

where Lemma 5.3.1 is used. We also observe that $T_3 \leqslant 0$.

We write our identity in the form

$$\frac{d}{ds}\Big(\int_{\mathcal{K}_{[s_0,s]}}\operatorname{div} W\,dxdt\Big) = T_1+T_2+T_3 \tag{5.3.3}$$

and obtain

$$\begin{aligned}
&\frac{d}{ds}\Big(\int_{\mathcal{K}_{[s_0,s_1]}}\operatorname{div} W\,dxdt\Big)\\
&\leqslant Cs^{-1}\Big\|\frac{ru\chi(r/t)}{s(1+r^2)^{1/2}}\Big\|_{L^2(\mathcal{H}_s)}\sum_a\big(\|(s/t)\partial_a\|_{L^2(\mathcal{H}_s)}+\|\underline{\partial}_a u\|_{L^2(\mathcal{H}_s)}\big).
\end{aligned}\tag{5.3.4}$$

On the other hand, we apply Stokes' formula in the region $\mathcal{K}_{[s_0,s_1]}$, and find

$$\begin{aligned}
&\int_{\mathcal{K}_{[s_0,s_1]}} \operatorname{div} W \, dxdt \\
&= \int_{\mathcal{H}_s} W \cdot n \, d\sigma + \int_{\mathcal{H}_{s_0}} W \cdot n \, d\sigma \\
&= \int_{\mathcal{H}_s} \frac{r^2}{1+r^2} \big|u\chi(r/t)s^{-1}\big|^2 dx - \int_{\mathcal{H}_{s_0}} \frac{r^2}{1+r^2} \big|u\chi(r/t)s^{-1}\big|^2 dx.
\end{aligned}$$

By differentiating this identity with respect to s, it follows that

$$\begin{aligned}
&\frac{d}{ds}\Big(\int_{\mathcal{K}_{[s_0,s_1]}} \operatorname{div} W \, dxdt\Big) = \frac{d}{ds}\Big(\int_{\mathcal{H}_s} \frac{r^2}{1+r^2}\big|u\chi(r/t)s^{-1}\big|^2 dx\Big) \\
&= 2\Big\|\frac{ru\chi(r/t)}{s(1+r^2)^{1/2}}\Big\|_{L^2(\mathcal{H}_s)} \frac{d}{ds}\Big\|\frac{ru\chi(r/t)}{s(1+r^2)^{1/2}}\Big\|_{L^2(\mathcal{H}_s)}.
\end{aligned} \tag{5.3.5}$$

Finally, combining (5.3.4) and (5.3.5) yields us

$$\frac{d}{ds}\Big\|\frac{ru\chi(r/t)}{s(1+r^2)^{1/2}}\Big\|_{L^2(\mathcal{H}_s)} \leqslant Cs^{-1}\sum_a \big(\|\underline{\partial}_a u\|_{L^2(\mathcal{H}_s)} + \|\underline{\partial}_a u\|_{L^2(\mathcal{H}_s)}\big)$$

and, by integration over the interval $[s_0, s]$,

$$\begin{aligned}
&\big\|r(1+r^2)^{-1/2}\chi(r/t)s^{-1}u\big\|_{L^2(\mathcal{H}_s)} \\
&\leqslant \big\|r(1+r^2)^{-1/2}\chi(r/t)s_0^{-1}u\big\|_{L^2(\mathcal{H}_{s_0})} \\
&\quad + C\sum_a \int_{s_0}^{s} \tau^{-1}\big(\|\underline{\partial}_a u\|_{L^2(\mathcal{H}_\tau)} + \|\underline{\partial}_a u\|_{L^2(\mathcal{H}_\tau)}\big).
\end{aligned} \tag{5.3.6}$$

In view of Lemma 5.3.1, we conclude that

$$\begin{aligned}
&\|\chi(r/t)s^{-1}u\|_{L^2(\mathcal{H}_s)} \\
&\leqslant C\,\|r(1+r^2)^{-1/2}\chi(r/t)s^{-1}u\|_{L^2(\mathcal{H}_s)} \\
&\leqslant C\,\big\|s_0^{-1}u\big\|_{L^2(\mathcal{H}_{s_0})} + C\sum_a \int_{s_0}^{s} \tau^{-1}\big(\|\underline{\partial}_a u\|_{L^2(\mathcal{H}_\tau)} + \|\underline{\partial}_a u\|_{L^2(\mathcal{H}_\tau)}\big).
\end{aligned} \tag{5.3.7}$$

The desired conclusion is reached by combining (5.3.2) with (5.3.7). □

Chapter 6

Revisiting scalar wave equations

6.1 Background and statement of the main result

In this chapter, we revisit the classical global existence theory for scalar nonlinear wave equations (with initial data imposed on a hyperboloid $\mathcal{H}_{s_0}$, with $s_0 \geqslant 1$):

$$\begin{aligned} \Box u &= P^{\alpha\beta}\partial_\alpha u \partial_\beta u \\ u|_{\mathcal{H}_{s_0}} &= u_0, \qquad \partial_t u|_{\mathcal{H}_{s_0}} = u_1, \end{aligned} \tag{6.1.1}$$

with smooth initial data u_0, u_1 compactly supported in the open ball $\mathcal{B}(0, s_0)$. We denote by $\mathcal{B}(0, s_0)$ the intersection of the spacelike hypersurface $\mathcal{H}_{s_0}$ and the cone $\mathcal{K} = \{(t,x) \,/\, |x| < t-1\}$. We impose the classical null condition on the bilinear form $P^{\alpha\beta}$, that is,

$$P^{\alpha\beta}\xi_\alpha\xi_\beta = 0 \quad \text{for all } \xi \in \mathbb{R}^4 \text{ satisfying } -\xi_0^2 + \sum_a \xi_a^2 = 0. \tag{6.1.2}$$

We are going to revisit this classical problem with the hyperboloidal foliation method and we establish that the energy of solutions is uniformly bounded, while, according to the classical technique of proof, the energy is only known to be at most polynomially increasing. The hyperboloidal energy $E_m = E_{m,0}$ was introduced in (2.1.11), while the admissible vector fields $Z \in \mathscr{Z}$ were defined in (2.1.7).

Theorem 6.1.1 (Existence theory for scalar wave equations). *There exist $\epsilon_0, C_1 > 0$ such that for all initial data satisfying*

$$E_m(s_0, Z^I u)^{1/2} \leqslant \epsilon \leqslant \epsilon_0 \qquad \text{for all } |I| \leqslant 3, \quad Z \in \mathscr{Z}, \tag{6.1.3}$$

the local-in-time solution u to the Cauchy problem (6.1.1) *extends to arbitrarily large times and, furthermore,*

$$E_m(s, Z^I u)^{1/2} \leqslant C_1\epsilon \qquad \text{for all } |I| \leqslant 3, \quad Z \in \mathscr{Z}, \tag{6.1.4}$$

and

$$\left|\partial_\alpha u(t,x)\right| \leqslant C_1 \varepsilon t^{-1}(t-|x|)^{-1/2}. \tag{6.1.5}$$

6.2 Structure of the proof

We proceed with a bootstrap strategy and, for some large constant $C_1 > 1$, we assume that, in a time interval $[s_0, s_1]$ the local solution satisfies the bound

$$\sum_{\substack{|I| \leqslant 3 \\ Z \in \mathscr{Z}}} E_m(s, Z^I u)^{1/2} \leqslant C_1 \epsilon \quad \text{for } s \in [s_0, s_1]. \tag{6.2.1}$$

We take

$$s_1 := \sup \Big\{ s \geqslant s_0 \, / \sum_{\substack{|I| \leqslant 3 \\ Z \in \mathscr{Z}}} E_m(\tau, Z^I u)^{1/2} \leqslant C_1 \epsilon \text{ for all } \tau \in [s_0, s] \Big\}$$

to be the largest such time and we suppose that it would be finite. Since $C_1 > 1$, by a continuity argument, we know that $s_1 > s_0$.

Our objective is to establish, for a suitable choice of $\epsilon_0, C_1 > 0$, that for all $\epsilon \leqslant \epsilon_0$,

$$\sum_{\substack{|I| \leqslant 3 \\ Z \in \mathscr{Z}}} E_m(s, Z^I u)^{1/2} \leqslant \frac{1}{2} C_1 \epsilon \quad \text{for } s \in [s_0, s_1]. \tag{6.2.2}$$

This leads us to

$$\sum_{\substack{|I| \leqslant 3 \\ Z \in \mathscr{Z}}} E_m(s_1, Z^I u)^{1/2} \leqslant \frac{1}{2} C_1 \epsilon,$$

and, by continuity,

$$s_1 < \sup \Big\{ s \geqslant s_0 \, / \sum_{\substack{|I| \leqslant 3 \\ Z \in \mathscr{Z}}} E_m(\tau, Z^I u) \leqslant C_1 \epsilon \text{ for all } \tau \in [s_0, s] \Big\}, \tag{6.2.3}$$

which would be a contradiction. We can then conclude that $s_1 = +\infty$. Namely, in view of the local-in-time existence theory (cf. Theorem 11.2.1), the solution u extends to all times.

In other words, our task reduces to proving the following result, and the rest of this chapter is devoted to its proof.

Proposition 6.2.1. *Let u be a solution to* (6.1.1) *defined in $[s_0, s_1]$ and with initial data satisfying*

$$\sum_{\substack{|I| \leqslant 3 \\ Z \in \mathscr{Z}}} E_m(s_0, Z^I u)^{1/2} \leqslant \varepsilon. \tag{6.2.4}$$

There exist constants $C_1, \varepsilon_0 > 0$ such that if u satisfies the estimate (6.2.1) *with $\varepsilon \leqslant \varepsilon_0$, then the estimate* (6.2.2) *holds.*

6.3 Energy estimate

The following lemma is essentially Proposition 2.3.1 in the special case of (6.1.1), but for convenience we provide here a direct proof.

Lemma 6.3.1. *If u is a solution to (6.1.1) defined in $[s_0, s_1]$, then the energy estimate*

$$\begin{aligned} &\sum_{\substack{|I|\leqslant 3\\ Z\in\mathscr{Z}}} E_m(s, Z^I u)^{1/2} \\ &\leqslant \sum_{\substack{|I|\leqslant 3\\ Z\in\mathscr{Z}}} E_m(s_0, Z^I u)^{1/2} + \int_{s_0}^{s} \sum_{\substack{|I|\leqslant 3\\ Z\in\mathscr{Z}}} \|Z^I(P^{\alpha\beta}\partial_\alpha u\partial_\beta u)\|_{L^2(\mathcal{H}_\tau)} d\tau. \end{aligned} \tag{6.3.1}$$

Proof. We apply to the equation (6.1.1) a product Z^I with $|I| \leqslant 3$ and, by recalling the commutation relation $[Z^I, \Box] = 0$, we find

$$\Box(Z^I u) = Z^I(P^{\alpha\beta}\partial_\alpha u\partial_\beta u).$$

By multiplying this equation by $\partial_t Z^I u$ and performing the standard energy calculation, we deduce that the function $\widetilde{u} := Z^I u$ satisfies

$$\frac{1}{2}\partial_t\Big((\partial_t\widetilde{u})^2 + \sum_a(\partial_a\widetilde{u})^2\Big) - \partial_a(\partial_a\widetilde{u}\partial_t\widetilde{u}) = Z^I(P^{\alpha\beta}\partial_\alpha\widetilde{u}\partial_\beta\widetilde{u})\partial_t\widetilde{u}.$$

Integrating this equation in the region $\mathcal{K}_{[s_0,s]}$, we have

$$\begin{aligned} &\int_{\mathcal{K}_{[s_0,s]}} \left(\frac{1}{2}\partial_t\Big((\partial_t\widetilde{u})^2 + \sum_a(\partial_a\widetilde{u})^2\Big) - \partial_a(\partial_a\widetilde{u}\partial_t\widetilde{u}\right) dtdx \\ &= \int_{\mathcal{K}_{[s_0,s]}} Z^I(P^{\alpha\beta}\partial_\alpha\widetilde{u}\partial_\beta u)\partial_t\widetilde{u}\, dtdx. \end{aligned} \tag{6.3.2}$$

By Huygens' principle, the solution is supported in the cone $\mathcal{K}$ and, in a neighborhood of the cone $\{|x| = t - 1\} \cap \{s_0 \leqslant \tau \leqslant s\}$, we have $\widetilde{u} = 0$. By Stokes' formula, the left-hand side of (6.3.2) reduces to

$$\begin{aligned} &\frac{1}{2}\int_{\mathcal{H}_s} \Big((\partial_t\widetilde{u})^2 + \sum_a(\partial_a\widetilde{u})^2, 2\partial_t\widetilde{u}\partial_a\widetilde{u}\Big)\cdot n\, d\sigma \\ &- \frac{1}{2}\int_{\mathcal{H}_{s_0}} (|\partial_t\widetilde{u}|^2 + \sum_a|\partial_a\widetilde{u}|^2, 2\partial_t\widetilde{u}\partial_a\widetilde{u}).n d\sigma, \end{aligned}$$

where n is the (future oriented) unit normal vector to the hyperboloids and $d\sigma$ is the induced Lebesgue measure on the hyperboloids, with

$$n = (t^2 + |x|^2)^{-1/2}(t, -x^a), \quad d\sigma = \frac{(t^2 + |x|^2)^{1/2}}{t}dx.$$

Hence, the left-hand side of (6.3.2) reads

$$\frac{1}{2}\int_{\mathcal{H}_s}\Big(|\partial_t\widetilde{u}|^2+\sum_a|\partial_a\widetilde{u}|^2+2\frac{x^a}{t}\partial_a\widetilde{u}\partial_t\widetilde{u}\Big)dx$$
$$-\frac{1}{2}\int_{\mathcal{H}_{s_0}}\Big(|\partial_t\widetilde{u}|^2+\sum_a|\partial_a\widetilde{u}|^2+2\frac{x^a}{t}\partial_a\widetilde{u}\partial_t\widetilde{u}\Big)dx,$$

which is $\frac{1}{2}E_m(s,Z^Iu)-\frac{1}{2}E_m(s_0,Z^Iu)$.

On the other hand, in the region $\mathcal{K}_{[s_0,s]}$, we use the change of variable $\tau=(t^2-|x|^2)^{1/2}$ and the identity $dtdx=(\tau/t)d\tau dx$, so that the right-hand side of (6.3.2) becomes

$$\int_{s_0}^{s}\int_{\mathcal{H}_\tau}(\tau/t)\partial_t u Z^I\big(P^{\alpha\beta}\partial_\alpha u\partial_\beta u\big)dxd\tau.$$

We thus conclude that (6.3.2) is equivalent to

$$\frac{1}{2}E_m(s,Z^Iu)-\frac{1}{2}E_m(s_0,Z^Iu)=\int_{s_0}^{s}ds\int_{\mathcal{H}_\tau}(\tau/t)\partial_t u Z^I\big(P^{\alpha\beta}\partial_\alpha u\partial_\beta u\big)dx. \tag{6.3.3}$$

Next, we differentiate (6.3.3) with respect to the variable s and obtain

$$\begin{aligned}E(\tau,Z^Iu)^{1/2}\frac{d}{d\tau}E(\tau,Z^Iu)^{1/2}&=\int_{\mathcal{H}_\tau}(\tau/t)\partial_t u Z^I\big(P^{\alpha\beta}\partial_\alpha u\partial_\beta u\big)dx\\&\leqslant\|(\tau/t)\partial_t u\|_{L^2(\mathcal{H}_\tau)}\big\|Z^I\big(P^{\alpha\beta}\partial_\alpha u\partial_\beta u\big)\big\|_{\mathcal{H}_\tau}.\end{aligned}$$

Recalling the expression of the hyperboloidal energy (2.3.2), we have $E(s,u)^{1/2}\geqslant\|(\tau/t)\partial_t u\|_{L^2(\mathcal{H}_s)}$ and therefore

$$\frac{d}{ds}\sum_{\substack{|I|\leqslant 3\\ Z\in\mathscr{Z}}}E(s,Z^Iu)^{1/2}\leqslant\sum_{\substack{|I|\leqslant 3\\ Z\in\mathscr{Z}}}\big\|Z^I\big(P^{\alpha\beta}\partial_\alpha u\partial_\beta u\big)\big\|_{\mathcal{H}_s}.$$

The conclusion follows by integrating over $[s_0,s]$. □

6.4 Basic L^2 and L^∞ estimates

In this section, from the bootstrap assumption (6.2.1), we deduce L^2 and L^∞ estimates. First of all, the following lemma is immediate in view of the expression of the hyperboloidal energy and (6.2.1).

Lemma 6.4.1 (Basic L^2 estimates). *By relying on (6.2.1), the following estimate hold for all $s\in[s_0,s_1]$:*

$$\sum_{\substack{|I|\leqslant 3\\ Z\in\mathscr{Z}}}\|\underline{\partial}_a Z^Iu\|_{L^2(\mathcal{H}_s)}\leqslant CC_1\epsilon, \tag{6.4.1a}$$

$$\sum_{\substack{|I|\leqslant 3\\ Z\in\mathscr{Z}}} \|(s/t)\underline{\partial}_0 Z^I u\|_{L^2(\mathcal{H}_s)} \leqslant CC_1\epsilon, \tag{6.4.1b}$$

where $C > 0$ is a (universal) constant.

Now, we combine Lemma 6.4.1 with the commutator estimates in Lemmas 3.3.1 and 3.3.2.

Lemma 6.4.2. *By relying on (6.2.1), the following estimate hold for all $s \in [s_0, s_1]$:*

$$\|Z^{I_1}\underline{\partial}_a Z^{I_2} u\|_{L^2(\mathcal{H}_s)} \leqslant CC_1\epsilon \quad \text{for all } |I_1| + |I_2| \leqslant 3, \tag{6.4.2a}$$

$$\|Z^{I_1}\big((s/t)\underline{\partial}_0 Z^{I_2} u\big)\|_{L^2(\mathcal{H}_s)} \leqslant CC_1\epsilon \quad \text{for all } |I_1| + |I_2| \leqslant 3, \tag{6.4.2b}$$

where $C > 0$ is a universal constant.

Furthermore, the following decay estimate is immediate in view of (6.4.2) and the Sobolev estimate on hyperboloids (5.1.1).

Lemma 6.4.3 (Basic L^∞ estimates). *By relying on (6.2.1), the following estimate hold for all $s \in [s_0, s_1]$:*

$$\|t^{3/2}\underline{\partial}_a Z^J u\|_{L^\infty(\mathcal{H}_s)} \leqslant CC_1\epsilon \quad \text{for all } |J| \leqslant 1, \tag{6.4.3a}$$

$$\|t^{1/2} s\underline{\partial}_0 Z^I u\|_{L^\infty(\mathcal{H}_s)} \leqslant CC_1\epsilon \quad \text{for all } |J| \leqslant 1, \tag{6.4.3b}$$

where $C > 0$ is a universal constant.

6.5 Estimate on the interaction term

We are now in a position to control the interaction term $P^{\alpha\beta}\partial_\alpha u \partial_\beta u$ with the help of the L^2 bound (6.4.2) and the L^∞ bound (6.4.3).

Lemma 6.5.1. *By reyling on the inequalities (6.4.2) and (6.4.3) and by assuming that the billinear form associated with $P^{\alpha\beta}$ is a null form, the following estimate holds for all $s \in [s_0, s_1]$:*

$$\big\|Z^I\big(P^{\alpha\beta}\partial_\alpha u \partial_\beta u\big)\big\|_{L^2(\mathcal{H}_s)} \leqslant CK\,(C_1\varepsilon)^2 s^{-3/2}, \tag{6.5.1}$$

where $C > 0$ is a universal constant and $K = \max_{\alpha,\beta}|P^{\alpha\beta}|$.

Proof. We recall Proposition 4.1.2 which tells us how to estimate a null form, that is

$$\begin{aligned}\left|Z^I\left(P^{\alpha\beta}\partial_\alpha u\partial_\beta u\right)\right| &\leqslant CK(s/t)^2\sum_{|I_1|+|I_2|\leqslant|I|}\left|Z^{I_1}\partial_t uZ^{I_2}\partial_t u\right| \\ &\quad+CK\sum_{\substack{a,\beta,\\|I_1|+|I_2|\leqslant|I|}}\Big(\left|Z^{I_1}\underline{\partial}_a u\,Z^{I_2}\underline{\partial}_\beta u\right|+\left|Z^{I_1}\underline{\partial}_\beta u\,Z^{I_2}\underline{\partial}_a u\right|\Big) \\ &=: T_1+T_2,\end{aligned}$$

where $K=\max_{\alpha,\beta}|P^{\alpha\beta}|$.

To estimate the L^2 norm of each term in T_1, we remark that $|I_1|+|I_2|\leqslant |I|\leqslant 3$ implies that $|I_1|\leqslant 1$ or $|I_2|\leqslant 1$. Without loss of generality, we assume $|I_2|\leqslant 1$ and then write

$$\begin{aligned}&\left\|(s/t)^2Z^{I_1}\partial_t uZ^{I_2}\partial_t u\right\|_{L^2(\mathcal{H}_s)} \\ &\leqslant \left\|(s/t)\partial_t u\right\|_{L^2(\mathcal{H}_s)}\left\|(s/t)t^{-1/2}s^{-1}\left(t^{1/2}sZ^{I_2}\partial_t u\right)\right\|_{L^\infty(\mathcal{H}_s)} \\ &\leqslant CC_1\epsilon\, s^{-3/2}CC_1\epsilon\leqslant C(C_1\epsilon)^2s^{-3/2}.\end{aligned}$$

The terms in T_2 are estimated along the same idea by writing, when $|I_1|\leqslant 1$,

$$\begin{aligned}\left\|Z^{I_1}\underline{\partial}_a u\,Z^{I_2}\underline{\partial}_\beta u\right\|_{L^2(\mathcal{H}_s)} &=\left\|t^{3/2}Z^{I_1}\underline{\partial}_a u\,t^{-3/2}(t/s)(s/t)Z^{I_2}\underline{\partial}_\beta u\right\|_{L^2(\mathcal{H}_s)} \\ &\leqslant\left\|t^{3/2}Z^{I_1}\underline{\partial}_a u\,s^{-3/2}(s/t)Z^{I_2}\underline{\partial}_\beta u\right\|_{L^2(\mathcal{H}_s)} \\ &=s^{-3/2}\left\|t^{3/2}Z^{I_1}\underline{\partial}_a u\right\|_{L^\infty(\mathcal{H}_s)}\left\|(s/t)Z^{I_2}\underline{\partial}_\beta u\right\|_{L^2(\mathcal{H}_s)} \\ &\leqslant C(C_1\epsilon)^2s^{-3/2}\end{aligned}$$

and, on the other hand when $|I_2|\leqslant 1$,

$$\begin{aligned}\left\|Z^{I_1}\underline{\partial}_a u\,Z^{I_2}\underline{\partial}_\beta u\right\|_{L^2(\mathcal{H}_s)} &=\left\|Z^{I_1}\underline{\partial}_a u\,t^{-1/2}s^{-1}\,t^{1/2}sZ^{I_2}\underline{\partial}_\beta u\right\|_{L^2(\mathcal{H}_s)} \\ &\leqslant\left\|Z^{I_1}\underline{\partial}_a u\,s^{-3/2}t^{1/2}sZ^{I_2}\underline{\partial}_\beta u\right\|_{L^2(\mathcal{H}_s)} \\ &=s^{-3/2}\left\|Z^{I_1}\underline{\partial}_a u\right\|_{L^2(\mathcal{H}_s)}\left\|t^{1/2}sZ^{I_2}\underline{\partial}_\beta u\right\|_{L^\infty(\mathcal{H}_s)} \\ &\leqslant C(C_1\epsilon)^2s^{-3/2}.\end{aligned}$$

□

6.6 Conclusion

Proof of Proposition 6.2.1. Now, we combine the null form estimate (6.5.1) and the energy estimate (6.3.1), and obtain

$$\begin{aligned}
\sum_{\substack{|I|\leqslant 3\\ Z\in\mathscr{Z}}} E_m(s, Z^I u)^{1/2} &\leqslant \sum_{\substack{|I|\leqslant 3\\ Z\in\mathscr{Z}}} E_m(s_0, Z^I u)^{1/2} + C(C_1\epsilon)^2 \int_{s_0}^{s} \tau^{-3/2}\, d\tau \\
&\leqslant \sum_{\substack{|I|\leqslant 3\\ Z\in\mathscr{Z}}} E_m(s_0, Z^I u)^{1/2} + C(C_1\epsilon)^2 \int_{1}^{+\infty} \tau^{-3/2}\, d\tau \\
&\leqslant \epsilon + C(C_1\epsilon)^2.
\end{aligned}$$

In order to conclude, we take $C_1 > 2$ and $\epsilon \leqslant \epsilon_0 = \frac{C_1-2}{2CC_1^2}$, and we find that

$$\sum_{\substack{|I|\leqslant 3\\ Z\in\mathscr{Z}}} E_m(s, Z^I u)^{1/2} \leqslant \frac{1}{2} C_1\epsilon.$$

□

The time-asymptotics of the solutions is also clear: (6.1.4) has already been proved by Proposition 6.2.1. To see (6.1.5), we remark that by Lemma 6.4.3,

$$|\partial_t u| \leqslant C_1\epsilon t^{-1/2} s^{-1} \leqslant C_1\epsilon t^{-1}(t-r)^{-1/2}, \quad |\underline{\partial}_a u| \leqslant C_1\epsilon t^{-3/2}.$$

By recalling the relation

$$\partial_a u = \underline{\partial}_a u - \frac{x^a}{t}\partial_t u,$$

then (6.1.5) follows. In view of the discussion at the beginning of Section 6.2, Theorem 6.1.1 is now established.

Chapter 7

Fundamental L^∞ and L^2 estimates

7.1 Objective of this chapter

We begin the discussion of the bootstrap arguments and we suppose that (2.4.5) holds on some time interval. Our aim is to derive additional estimates from these assumptions. In the present chapter, we are able to deduce several L^2 and L^∞ estimates on the solution and its derivatives. These rather immediate estimates will serve as a basis for the following chapters. The estimates in this chapter are classified into two groups: L^2 estimates and L^∞ estimates:

- The L^2-type estimates themselves are classified into two generations:
  - The estimates of the first generation are immediate consequences of (2.4.5).
  - The estimates of the second generation are deduced from those of the first generation, by recalling the commutator estimates (cf. Lemmas 3.3.1 and 3.3.2). These are the L^2 bounds that will be more often used in the following discussion.
- The L^∞-type estimates are also classified into two generations:
  - The estimates of the first generation follow immediately from (2.4.5) and the Sobolev inequalities (5.1.1).
  - The estimates of the second generation are deduced from the ones of the first generation and the commutator estimates in Lemmas 3.3.1 and 3.3.2.

In the following, the letter C will be used to represent a constant who depends on the structure of the system (1.2.1), such as the number of equations n_0, the number of wave components j_0, and the Klein-Gordon constants $c_{\check{i}}$.

7.2 L^2 estimates of the first generation

From the expression of the energy (2.3.2) and the energy assumption (2.4.5), the following estimates hold of all $|I^\sharp| \leqslant 5$ and $|I^\dagger| \leqslant 4$:

$$\sum_{\alpha,i} \|(s/t)\partial_\alpha Z^{I^\sharp} w_i\|_{L^2(\mathcal{H}_s)} + \sum_{\alpha,i} \|(s/t)\underline{\partial}_\alpha Z^{I^\sharp} w_i\|_{L^2(\mathcal{H}_s)} \leqslant CC_1\epsilon s^\delta, \tag{7.2.1a}$$

$$\sum_{\alpha,i} \|(s/t)\partial_\alpha Z^{I^\dagger} w_i\|_{L^2(\mathcal{H}_s)} + \sum_{\alpha,i} \|(s/t)\underline{\partial}_\alpha Z^{I^\dagger} w_i\|_{L^2(\mathcal{H}_s)} \leqslant CC_1\epsilon s^{\delta/2}, \tag{7.2.1b}$$

$$\sum_{a,i} \|\underline{\partial}_a Z^{I^\sharp} w_i\|_{L^2(\mathcal{H}_s)} \leqslant CC_1\epsilon s^\delta, \tag{7.2.1c}$$

$$\sum_{a,i} \|\underline{\partial}_a Z^{I^\dagger} w_i\|_{L^2(\mathcal{H}_s)} \leqslant CC_1\epsilon s^{\delta/2}, \tag{7.2.1d}$$

$$\sum_{\check{i}} \|Z^{I^\sharp} v_{\check{i}}\|_{L^2(\mathcal{H}_s)} \leqslant CC_1\epsilon s^\delta, \tag{7.2.1e}$$

$$\sum_{\check{i}} \|Z^{I^\dagger} v_{\check{i}}\|_{L^2(\mathcal{H}_s)} \leqslant CC_1\epsilon s^{\delta/2}. \tag{7.2.1f}$$

By taking

$$\begin{aligned} &Z^{I^\sharp} = \partial_\alpha Z^{I^\dagger}, \quad Z^{I^\dagger} = \partial_\alpha Z^I, \\ &Z^{I^\sharp} = t\underline{\partial}_a Z^{I^\dagger} = L_a Z^{I^\dagger}, \quad Z^{I^\dagger} = t\underline{\partial}_a Z^I = L_a Z^I \end{aligned}$$

in (7.2.1e) and (7.2.1f), we have especially the following estimates on Klein-Gordon components for $|I^\dagger| \leqslant 4$ and $|I| \leqslant 3$:

$$\sum_{\check{i},\alpha} \|\partial_\alpha Z^{I^\dagger} v_{\check{i}}\|_{L^2(\mathcal{H}_s)} + \sum_{\check{i},a} \|\underline{\partial}_\alpha Z^{I^\dagger} v_{\check{i}}\|_{L^2(\mathcal{H}_s)} \leqslant CC_1\epsilon s^\delta, \tag{7.2.2a}$$

$$\sum_{\check{i},\alpha} \|\partial_\alpha Z^I v_{\check{i}}\|_{L^2(\mathcal{H}_s)} + \sum_{\check{i},a} \|\underline{\partial}_\alpha Z^I v_{\check{i}}\|_{L^2(\mathcal{H}_s)} \leqslant CC_1\epsilon s^{\delta/2}, \tag{7.2.2b}$$

$$\sum_{\check{i},a} \|t\underline{\partial}_a Z^{I^\dagger} v_{\check{i}}\|_{L^2(\mathcal{H}_s)} \leqslant CC_1\epsilon s^\delta, \tag{7.2.2c}$$

$$\sum_{\check{\imath},a} \|t\underline{\partial}_a Z^I v_{\check{\imath}}\|_{L^2(\mathcal{H}_s)} \leqslant CC_1\epsilon s^{\delta/2}. \tag{7.2.2d}$$

The bound on $\underline{\partial}_\alpha Z^{I^\dagger} v_{\check{\imath}}$ in (7.2.2a) is derived from the estimate on $\partial_t Z^{I^\dagger} v_{\check{\imath}}$ and the estimates on $\underline{\partial}_a Z^{I^\dagger} v_{\check{\imath}}$.

For any $|I| \leqslant 3$, from (2.3.2) and (2.4.5e) we control the wave components:

$$\sum_{\alpha,\hat{\imath}} \|(s/t)\partial_\alpha Z^I u_{\hat{\imath}}\|_{L^2(\mathcal{H}_s)} + \sum_{\alpha,\hat{\imath}} \|(s/t)\underline{\partial}_\alpha Z^I u_{\hat{\imath}}\|_{L^2(\mathcal{H}_s)} \leqslant CC_1\epsilon, \tag{7.2.3a}$$

$$\sum_{a,\hat{\imath}} \|\underline{\partial}_a Z^I u_{\hat{\imath}}\|_{L^2(\mathcal{H}_s)} \leqslant CC_1\epsilon. \tag{7.2.3b}$$

7.3 L^2 estimates of the second generation

By using Lemma 3.3.2, we can commute the vector fields under consideration and, relying on the estimates established in the previous section, we obtain the following result.

The first group of estimates is obtained by (3.3.1), (3.3.2), and (7.2.1):

$$\sum_{\hat{\imath},\alpha} \|(s/t) Z^{I^\sharp}\partial_\alpha w_i\|_{L^2(\mathcal{H}_s)} + \sum_{\hat{\imath},\alpha} \|(s/t) Z^{I^\sharp}\underline{\partial}_\alpha w_i\|_{L^2(\mathcal{H}_s)} \leqslant CC_1\epsilon s^\delta, \tag{7.3.1a}$$

$$\sum_{\hat{\imath},\alpha} \|(s/t) Z^{I^\dagger}\partial_\alpha w_i\|_{L^2(\mathcal{H}_s)} + \sum_{\hat{\imath},\alpha} \|(s/t) Z^{I^\dagger}\underline{\partial}_\alpha w_i\|_{L^2(\mathcal{H}_s)} \leqslant CC_1\epsilon s^{\delta/2}, \tag{7.3.1b}$$

$$\sum_{\hat{\imath},a} \|Z^{I^\sharp}\underline{\partial}_a w_{\hat{\imath}}\|_{L^2(\mathcal{H}_s)} \leqslant CC_1\epsilon s^\delta, \tag{7.3.1c}$$

$$\sum_{\hat{\imath},a} \|Z^{I^\dagger}\underline{\partial}_a w_{\hat{\imath}}\|_{L^2(\mathcal{H}_s)} \leqslant CC_1\epsilon s^{\delta/2}, \tag{7.3.1d}$$

$$\sum_{\check{\jmath}} \|Z^{I^\sharp} v_{\check{\jmath}}\|_{L^2(\mathcal{H}_s)} \leqslant CC_1\epsilon s^\delta, \tag{7.3.1e}$$

$$\sum_{\check{\jmath}} \|Z^{I^\dagger} v_{\check{\jmath}}\|_{L^2(\mathcal{H}_s)} \leqslant CC_1\epsilon s^{\delta/2}. \tag{7.3.1f}$$

The second group of estimates follows from (3.3.1), (3.3.2), and (7.2.2):

$$\sum_{\check{\imath},\alpha} \|Z^{I^\dagger}\partial_\alpha v_{\check{\imath}}\|_{L^2(\mathcal{H}_s)} + \sum_{\check{\imath},\alpha} \|Z^{I^\dagger}\underline{\partial}_\alpha v_{\check{\imath}}\|_{L^2(\mathcal{H}_s)} \leqslant CC_1\epsilon s^\delta, \tag{7.3.2a}$$

$$\sum_{\check{\imath},\alpha} \|Z^I \partial_\alpha v_{\check{\imath}}\|_{L^2(\mathcal{H}_s)} + \sum_{\check{\imath},\alpha} \|Z^I \underline{\partial}_\alpha v_{\check{\imath}}\|_{L^2(\mathcal{H}_s)} \leqslant CC_1\epsilon s^{\delta/2}, \tag{7.3.2b}$$

$$\sum_{\check{\imath},a} \|tZ^{I^\dagger} \underline{\partial}_a v_{\check{\imath}}\|_{L^2(\mathcal{H}_s)} \leqslant CC_1\epsilon s^{\delta}, \tag{7.3.2c}$$

$$\sum_{\check{\imath},a} \|tZ^{I} \underline{\partial}_a v_{\check{\imath}}\|_{L^2(\mathcal{H}_s)} \leqslant CC_1\epsilon s^{\delta/2}. \tag{7.3.2d}$$

The third group of estimates follows from (3.3.1), (3.3.2), and (7.2.3):

$$\sum_{\hat{\imath},\alpha} \|(s/t)Z^I \partial_\alpha u_{\hat{\imath}}\|_{L^2(\mathcal{H}_s)} + \sum_{\hat{\imath},\alpha} \|(s/t)Z^I \underline{\partial}_\alpha u_{\hat{\imath}}\|_{L^2(\mathcal{H}_s)} \leqslant CC_1\epsilon, \tag{7.3.3a}$$

$$\sum_{\hat{\imath},a} \|Z^I \underline{\partial}_a u_{\hat{\imath}}\|_{L^2(\mathcal{H}_s)} \leqslant CC_1\epsilon. \tag{7.3.3b}$$

The fourth group of estimates concerns the second order derivatives. The first is deduced from (5.2.5), (2.4.5a), and (2.4.5b), while the second is deduced from (5.2.5), (2.4.5d), and (2.4.5c), and the last one from (5.2.5) and (2.4.5e):

$$\sum_{a,\beta,i} \|s\underline{\partial}_a\underline{\partial}_\beta Z^{I^\dagger} w_i\|_{L^2(\mathcal{H}_s)} + \sum_{a,\beta,i} \|s\underline{\partial}_\beta\underline{\partial}_a Z^{I^\dagger} w_i\|_{L^2(\mathcal{H}_s)} \leqslant CC_1\epsilon s^{\delta}, \tag{7.3.4a}$$

$$\sum_{a,\beta,i} \|s\underline{\partial}_a\underline{\partial}_\beta Z^{I} w_i\|_{L^2(\mathcal{H}_s)} + \sum_{a,\beta,i} \|s\underline{\partial}_\beta\underline{\partial}_a Z^{I} w_i\|_{L^2(\mathcal{H}_s)} \leqslant CC_1\epsilon s^{\delta/2}, \tag{7.3.4b}$$

$$\sum_{a,\beta,\hat{\imath}} \|s\underline{\partial}_a\underline{\partial}_\beta Z^{I^\flat} u_{\hat{\imath}}\|_{L^2(\mathcal{H}_s)} + \sum_{a,\beta,\hat{\imath}} \|s\underline{\partial}_\beta\underline{\partial}_a Z^{I^\flat} u_{\hat{\imath}}\|_{L^2(\mathcal{H}_s)} \leqslant CC_1\epsilon, \tag{7.3.4c}$$

where the order of $I^\flat$ is less or equal to 2. We can also use the commutator estimates (3.3.4) and obtain

$$\sum_{a,\beta,i} \|sZ^{I^\dagger}\underline{\partial}_a\underline{\partial}_\beta w_i\|_{L^2(\mathcal{H}_s)} + \sum_{a,\beta,i} \|sZ^{I^\dagger}\underline{\partial}_\beta\underline{\partial}_a w_i\|_{L^2(\mathcal{H}_s)} \leqslant CC_1\epsilon s^{\delta}, \tag{7.3.5a}$$

$$\sum_{a,\beta,i} \|sZ^{I}\underline{\partial}_a\underline{\partial}_\beta w_i\|_{L^2(\mathcal{H}_s)} + \sum_{a,\beta,i} \|sZ^{I}\underline{\partial}_\beta\underline{\partial}_a w_i\|_{L^2(\mathcal{H}_s)} \leqslant CC_1\epsilon s^{\delta/2}, \tag{7.3.5b}$$

$$\sum_{a,\beta,\hat{\imath}} \|sZ^{I^\flat}\underline{\partial}_a\underline{\partial}_\beta u_{\hat{\imath}}\|_{\mathcal{H}_s} + \sum_{a,\beta,\hat{\imath}} \|sZ^{I^\flat}\underline{\partial}_\beta\underline{\partial}_a u_{\hat{\imath}}\|_{\mathcal{H}_s} \leqslant CC_1\epsilon. \tag{7.3.5c}$$

Finally, we have the L^2 estimates for the wave components themselves* for all $|I^\sharp| \leqslant 5$ and all $|I^\dagger| \leqslant 4$:

$$\left\| s^{-1} Z^{I^\sharp} u_{\hat{\imath}} \right\|_{L^2(\mathcal{H}_s)} \leqslant CC_1 \epsilon s^\delta, \tag{7.3.6a}$$

$$\left\| s^{-1} Z^{I^\dagger} u_{\hat{\imath}} \right\|_{L^2(\mathcal{H}_s)} \leqslant CC_1 \epsilon s^{\delta/2}, \tag{7.3.6b}$$

$$\left\| t^{-1} Z^{I} u_{\hat{\imath}} \right\|_{L^2(\mathcal{H}_s)} \leqslant CC_1 \epsilon. \tag{7.3.6c}$$

The first two inequalities are direct consequences of (7.2.1a), (7.2.1b), (7.2.1c), and (7.2.1d) combined with Proposition 5.3.1. The last inequality is a result of Lemma 5.3.1 combined with (7.2.3b).

7.4 L^∞ estimates of the first generation

For convenience in the presentation, we introduce a new convention (which is parallel to the index convention already made in (2.4.1)):

$$\begin{aligned} J^\sharp &\quad \text{index of order} \leqslant 3, \\ J^\dagger &\quad \text{index of order} \leqslant 2, \\ J &\quad \text{index of order} \leqslant 1. \end{aligned} \tag{7.4.1}$$

Now we give the decay estimates based on the energy assumption (2.4.5) and the Sobolev-type inequality given in Proposition 5.1.1.

The first group of inequalities is a direct consequence of (5.2.1), (2.4.5b), and (2.4.5a):

$$\sup_{\mathcal{H}_s} \left(s t^{1/2} |\partial_\alpha Z^{J^\sharp} w_j| \right) + \sup_{\mathcal{H}_s} \left(s t^{1/2} |\underline{\partial}_\alpha Z^{J^\sharp} w_j| \right) \leqslant CC_1 \epsilon s^\delta, \tag{7.4.2a}$$

$$\sup_{\mathcal{H}_s} \left(s t^{1/2} |\partial_\alpha Z^{J^\dagger} w_j| \right) + \sup_{\mathcal{H}_s} \left(s t^{1/2} |\underline{\partial}_\alpha Z^{J^\dagger} w_j| \right) \leqslant CC_1 \epsilon s^{\delta/2}, \tag{7.4.2b}$$

$$\sup_{\mathcal{H}_s} \left(t^{3/2} |\underline{\partial}_a Z^{J^\sharp} w_j| \right) \leqslant CC_1 \epsilon s^\delta, \tag{7.4.2c}$$

$$\sup_{\mathcal{H}_s} \left(t^{3/2} |\underline{\partial}_a Z^{J^\dagger} w_j| \right) \leqslant CC_1 \epsilon s^{\delta/2}, \tag{7.4.2d}$$

*In (7.3.6a), if we want to be more precise, $C = C'(C_0/C_1 + 1 + \delta^{-1})$ where C' is a constant depending only on the structure of the system. We see that it also depends on C_0 and δ. But as we can assume that $1/12 \leqslant \delta < 1/6$ and $C_1 \geqslant C_0$, we denote it again by C and regard it as a constant determined by the structure of the system.

$$\sup_{\mathcal{H}_s} \left(t^{3/2}|Z^{J^\sharp} v_{\check{j}}|\right) \leqslant CC_1\epsilon s^\delta, \tag{7.4.2e}$$

$$\sup_{\mathcal{H}_s} \left(t^{3/2}|Z^{J^\dagger} v_{\check{j}}|\right) \leqslant CC_1\epsilon s^{\delta/2}. \tag{7.4.2f}$$

The second group of estimates is a special case of the second estimate, by taking

$$Z^{J^\sharp} = \partial_\alpha Z^{J^\dagger}, \quad Z^{J^\sharp} = L_a Z^{J^\dagger} = t\underline{\partial}_a Z^{J^\dagger},$$
$$Z^{J^\dagger} = \partial_\alpha Z^J, \quad Z^{J^\dagger} = L_a Z^J = t\underline{\partial}_a Z^J$$

in (7.4.2e) and (7.4.2f):

$$\sup_{\mathcal{H}_s} \left(t^{3/2}|\partial_\alpha Z^{J^\dagger} v_{\check{j}}|\right) + \sup_{\mathcal{H}_s} \left(t^{3/2}|\underline{\partial}_\alpha Z^{J^\dagger} v_{\check{j}}|\right) \leqslant CC_1\epsilon s^\delta, \tag{7.4.3a}$$

$$\sup_{\mathcal{H}_s} \left(t^{3/2}|\partial_\alpha Z^J v_{\check{j}}|\right) + \sup_{\mathcal{H}_s} \left(t^{3/2}|\underline{\partial}_\alpha Z^J v_{\check{j}}|\right) \leqslant CC_1\epsilon s^{\delta/2}, \tag{7.4.3b}$$

$$\sup_{\mathcal{H}_s} \left(t^{5/2}|\underline{\partial}_a Z^{J^\dagger} v_{\check{j}}|\right) \leqslant CC_1\epsilon s^\delta, \tag{7.4.3c}$$

$$\sup_{\mathcal{H}_s} \left(t^{5/2}|\underline{\partial}_a Z^J v_{\check{j}}|\right) \leqslant CC_1\epsilon s^{\delta/2}. \tag{7.4.3d}$$

The estimates in the third group follow from (5.2.1) (a) and (b) combined with (2.4.5e):

$$\sup_{\mathcal{H}_s} \left(st^{1/2}|\partial_\alpha Z^J u_{\hat{k}}|\right) + \sup_{\mathcal{H}_s} \left(st^{1/2}|\underline{\partial}_\alpha Z^J u_{\hat{k}}|\right) \leqslant CC_1\epsilon, \tag{7.4.4a}$$

$$\sup_{\mathcal{H}_s} \left(t^{3/2}|\underline{\partial}_a Z^J u_{\hat{k}}|\right) \leqslant CC_1\epsilon. \tag{7.4.4b}$$

The fourth group concerns the "second order derivative" of the solution. They are deduced from (5.2.4) and (2.4.5):

$$\sup_{\mathcal{H}_s} \left(t^{3/2}s|\underline{\partial}_\alpha\underline{\partial}_a Z^{J^\dagger} w_j|\right) + \sup_{\mathcal{H}_s} \left(t^{3/2}s|\underline{\partial}_a\underline{\partial}_\alpha Z^{J^\dagger} w_j|\right) \leqslant CC_1\epsilon s^\delta, \tag{7.4.5a}$$

$$\sup_{\mathcal{H}_s} \left(t^{3/2}s|\underline{\partial}_\alpha\underline{\partial}_a Z^J w_j|\right) + \sup_{\mathcal{H}_s} \left(t^{3/2}s|\underline{\partial}_a\underline{\partial}_\alpha Z^J w_j|\right) \leqslant CC_1\epsilon s^{\delta/2}, \tag{7.4.5b}$$

$$\sup_{\mathcal{H}_s} \left(t^{3/2}s|\underline{\partial}_\alpha\underline{\partial}_a u_{\hat{\jmath}}|\right) + \sup_{\mathcal{H}_s} \left(t^{3/2}s|\underline{\partial}_a\underline{\partial}_\alpha u_{\hat{\jmath}}|\right) \leqslant CC_1\epsilon. \tag{7.4.5c}$$

7.5 L^∞ estimates of the second generation

The estimates in this section follow from the first generation L^∞ estimates established in the previous section, combined with the commutator estimates in Lemmas 3.3.1 and 3.3.2.

The first group of estimates are results of (3.3.1), (3.3.2) combined with (7.4.2):

$$\sup_{\mathcal{H}_s} \left|st^{1/2} Z^{J^\sharp} \partial_\alpha w_j\right| + \sup_{\mathcal{H}_s} \left|st^{1/2} Z^{J^\sharp} \underline{\partial}_\alpha w_j\right| \leqslant CC_1 \epsilon s^\delta, \tag{7.5.1a}$$

$$\sup_{\mathcal{H}_s} \left|st^{1/2} Z^{J^\dagger} \partial_\alpha w_j\right| + \sup_{\mathcal{H}_s} \left|st^{1/2} Z^{J^\dagger} \underline{\partial}_\alpha w_j\right| \leqslant CC_1 \epsilon s^{\delta/2}, \tag{7.5.1b}$$

$$\sup_{\mathcal{H}_s} \left|t^{3/2} Z^{J^\sharp} \underline{\partial}_a w_j\right| \leqslant CC_1 \epsilon s^\delta, \tag{7.5.1c}$$

$$\sup_{\mathcal{H}_s} \left|t^{3/2} Z^{J^\dagger} \underline{\partial}_a w_j\right| \leqslant CC_1 \epsilon s^{\delta/2}, \tag{7.5.1d}$$

$$\sup_{\mathcal{H}_s} \left|t^{3/2} Z^{J^\sharp} v_{\check{k}}\right| \leqslant CC_1 \epsilon s^\delta, \tag{7.5.1e}$$

$$\sup_{\mathcal{H}_s} \left|t^{3/2} Z^{J^\dagger} v_{\check{k}}\right| \leqslant CC_1 \epsilon s^{\delta/2}. \tag{7.5.1f}$$

The second group consists of the following estimates. They are results of (3.3.1), (3.3.2) combined with (7.4.3):

$$\sup_{\mathcal{H}_s} \left|t^{3/2} Z^{J^\dagger} \partial_\alpha v_{\check{\jmath}}\right| + \sup_{\mathcal{H}_s} \left|t^{3/2} Z^{J^\dagger} \underline{\partial}_\alpha v_{\check{\jmath}}\right| \leqslant CC_1 \epsilon s^\delta, \tag{7.5.2a}$$

$$\sup_{\mathcal{H}_s} \left|t^{3/2} Z^{J} \partial_\alpha v_{\check{\jmath}}\right| + \sup_{\mathcal{H}_s} \left|t^{3/2} Z^{J} \underline{\partial}_\alpha v_{\check{\jmath}}\right| \leqslant CC_1 \epsilon s^{\delta/2}, \tag{7.5.2b}$$

$$\sup_{\mathcal{H}_s} \left|t^{5/2} Z^{J^\dagger} \underline{\partial}_a v_{\check{\jmath}}\right| \leqslant CC_1 \epsilon s^\delta, \tag{7.5.2c}$$

$$\sup_{\mathcal{H}_s} \left|t^{5/2} Z^{J} \underline{\partial}_a v_{\check{\jmath}}\right| \leqslant CC_1 \epsilon s^{\delta/2}. \tag{7.5.2d}$$

The third group consists of the following estimates which are direct consequences of (3.3.1) and (3.3.2) combined with (7.4.4):

$$\sup_{\mathcal{H}_s} \left|st^{1/2} Z^{J} \partial_\alpha u_{\hat{\jmath}}\right| + \sup_{\mathcal{H}_s} \left|st^{1/2} Z^{J} \underline{\partial}_\alpha u_{\hat{\jmath}}\right| \leqslant CC_1 \epsilon, \tag{7.5.3a}$$

$$\sup_{\mathcal{H}_s} \left|t^{3/2} Z^J \underline{\partial}_a u_{\hat{\jmath}}\right| \leqslant CC_1 \epsilon. \tag{7.5.3b}$$

The estimates of the fourth group are deduced from (3.3.4), (7.4.5a), (7.4.5b), and (7.4.5c):

$$\sup_{\mathcal{H}_s} \left(t^{3/2} s Z^{J^\dagger} \underline{\partial}_\alpha \underline{\partial}_a w_{\check{\jmath}}\right) + \sup_{\mathcal{H}_s} \left(t^{3/2} s Z^{J^\dagger} \underline{\partial}_a \underline{\partial}_\alpha w_{\check{\jmath}}\right) \leqslant CC_1 \epsilon s^\delta, \tag{7.5.4a}$$

$$\sup_{\mathcal{H}_s} \left(t^{3/2} s Z^J \underline{\partial}_\alpha \underline{\partial}_a w_{\check{\jmath}}\right) + \sup_{\mathcal{H}_s} \left(t^{3/2} s Z^J \underline{\partial}_a \underline{\partial}_\alpha w_{\check{\jmath}}\right) \leqslant CC_1 \epsilon s^{\delta/2}, \tag{7.5.4b}$$

$$\sup_{\mathcal{H}_s} \left|s t^{3/2} \underline{\partial}_a \underline{\partial}_\beta u_{\hat{\jmath}}\right| + \sup_{\mathcal{H}_s} \left|s t^{3/2} \underline{\partial}_\alpha \underline{\partial}_b u_{\hat{\jmath}}\right| \leqslant CC_1 \epsilon s. \tag{7.5.4c}$$

We finally can write down the decay estimates for the wave components, which follow from Lemma 5.2.3 combined with (2.4.5):

$$\sup_{\mathcal{H}_s} \left(t^{3/2} s^{-1} |Z^{J^\sharp} u_{\hat{\imath}}|\right) \leqslant CC_1 \epsilon s^\delta, \tag{7.5.5a}$$

$$\sup_{\mathcal{H}_s} \left(t^{3/2} s^{-1} |Z^{J^\dagger} u_{\hat{\imath}}|\right) \leqslant CC_1 \epsilon s^{\delta/2}, \tag{7.5.5b}$$

$$\sup_{\mathcal{H}_s} \left(t^{3/2} s^{-1} |Z^J u_{\hat{\imath}}|\right) \leqslant CC_1 \epsilon. \tag{7.5.5c}$$

Chapter 8

Second-order derivatives of the wave components

8.1 Preliminaries

The estimates in this chapter concern "second-order" terms $\partial_\alpha \partial_\beta Z^I u_{\hat{\imath}}$ and $Z^I \partial_\alpha \partial_\beta u_{\hat{\imath}}$, which will also be used for the control of

$$u_{\hat{\imath}} \partial_\alpha \partial_\beta u_{\hat{\jmath}}.$$

This chapter is more technical than the derivation of our earlier estimates and our strategy now is as follows. We are going to analyze the structure of certain second-order derivatives of the wave components in the semi-hyperboloidal frame and concentrate first on the component of the Hessian $\underline{\partial}_0 \underline{\partial}_0 Z^I u_{\hat{\imath}}$ or, equivalently, $\partial_t \partial_t Z^I u_{\hat{\imath}}$. Other components have a main part which can be expressed in terms of this component. Next, we will analyze $\partial_t \partial_t Z^I u_{\hat{\imath}}$ and give a general sketch of the proof, while postponing the technical aspects to the last two sections.

More precisely, let us first reduce the derivation of an estimating of the Hessian of $Z^I u_{\hat{\imath}}$ to an estimate of its component $\underline{\partial}_0 \underline{\partial}_0 Z^I u_{\hat{\imath}}$.

Lemma 8.1.1. *If u is a smooth function compactly supported in the half-cone $\mathcal{K} = \{|x| \leqslant t - 1\}$, then for all index I one has*

$$|\partial_\alpha \partial_\beta Z^I u| \leqslant |\partial_t \partial_t Z^I u| + C \sum_{a,\beta} |\underline{\partial}_a \partial_\beta Z^I u| + \frac{C}{t} \sum_\gamma |\partial_\gamma Z^I u|,$$

where C is a universal constant.

Proof. We recall the following identity based on a change of frame (between the natural frame and the semi-hyperboloidal frame):

$$\begin{aligned}\partial_\alpha\partial_\beta Z^I u =& \Psi_\alpha^{\alpha'}\Psi_\beta^{\beta'}\underline{\partial}_{\alpha'}\underline{\partial}_{\beta'}Z^I u + \partial_\alpha\big(\Psi_\beta^{\beta'}\big)\underline{\partial}_\beta Z^I u\\ =&\Psi_\alpha^0\Psi_\beta^0\underline{\partial}_0\underline{\partial}_0 Z^I u + \sum_b \Psi_\alpha^0\Psi_\beta^b\underline{\partial}_0\underline{\partial}_b Z^I u\\ &+ \sum_a \Psi_\alpha^a\Psi_\beta^0\underline{\partial}_a\underline{\partial}_0 Z^I u + \sum_{a,b}\Psi_\alpha^a\Psi_\beta^b\underline{\partial}_a\underline{\partial}_b Z^I u + \partial_\alpha\big(\Psi_\beta^{\beta'}\big)\underline{\partial}_\beta Z^I u.\end{aligned}$$

By observing that $\big|\partial_\alpha\Psi_\beta^{\beta'}\big| \leqslant \frac{C}{t}$ and

$$\big|[\underline{\partial}_b, \partial_t](Z^I u)\big| \leqslant Ct^{-1}\sum_\alpha|\partial_\alpha Z^I u|,$$

we can write

$$\sum_b|\partial_t\underline{\partial}_b Z^I u| + \sum_{a,b}|\underline{\partial}_a\underline{\partial}_b Z^I u| \leqslant \sum_{a,\beta}|\underline{\partial}_a\partial_\beta Z^I u| + Ct^{-1}\sum_\gamma|\partial_\gamma Z^I u|.$$

□

Now we assume that the energy is controlled.

Lemma 8.1.2. *Under the energy assumption (2.4.5), the following L^2 estimates hold for all $|I^\dagger| \leqslant 4, |I| \leqslant 3$ and $|I^\flat| \leqslant 2$:*

$$\big\|s^3t^{-2}\partial_\alpha\partial_\beta Z^{I^\dagger}u_{\widehat{\imath}}\big\|_{L^2(\mathcal{H}_s)} \leqslant \big\|s^3t^{-2}\partial_t\partial_t Z^{I^\dagger}u_{\widehat{\imath}}\big\|_{L^2(\mathcal{H}_s)} + CC_1\epsilon s^\delta, \tag{8.1.1a}$$

$$\big\|s^3t^{-2}\partial_\alpha\partial_\beta Z^{I}u_{\widehat{\imath}}\big\|_{L^2(\mathcal{H}_s)} \leqslant \big\|s^3t^{-2}\partial_t\partial_t Z^{I}u_{\widehat{\imath}}\big\|_{L^2(\mathcal{H}_s)} + CC_1\epsilon s^{\delta/2}, \tag{8.1.1b}$$

$$\big\|s^3t^{-2}\partial_\alpha\partial_\beta Z^{I^\flat}u_{\widehat{\imath}}\big\|_{L^2(\mathcal{H}_s)} \leqslant \big\|s^3t^{-2}\partial_t\partial_t Z^{I^\flat}u_{\widehat{\imath}}\big\|_{L^2(\mathcal{H}_s)} + CC_1\epsilon. \tag{8.1.1c}$$

Furthermore, the following L^∞ estimates hold for all $|J^\dagger| \leqslant 2$ and $|J| \leqslant 1$:

$$\sup_{\mathcal{H}_s}|s^3t^{-1/2}\partial_\alpha\partial_\beta Z^{J^\dagger}u| \leqslant \sup_{\mathcal{H}_s}|s^3t^{-1/2}\partial_t\partial_t Z^{J^\dagger}u| + CC_1\epsilon s^\delta, \tag{8.1.2a}$$

$$\sup_{\mathcal{H}_s}|s^3t^{-1/2}\partial_\alpha\partial_\beta Z^{J}u| \leqslant \sup_{\mathcal{H}_s}|s^3t^{-1/2}\partial_t\partial_t Z^{J}u| + CC_1\epsilon s^{\delta/2}, \tag{8.1.2b}$$

$$\sup_{\mathcal{H}_s}|s^3t^{-1/2}\partial_\alpha\partial_\beta u| \leqslant \sup_{\mathcal{H}_s}|s^3t^{-1/2}\partial_t\partial_t u| + CC_1\epsilon. \tag{8.1.2c}$$

Proof. The estimate (8.1.1a) is a combination of Lemma 8.1.1 and the estimate (7.3.4a), while (8.1.1b) is a combination of Lemma 8.1.1 and the estimate (7.3.4b). On the other hand, (8.1.1c) is a combination of Lemma 8.1.1 and the estimate (7.3.4c). For the L^∞ estimates, we combine Lemma 8.1.1 with (7.4.5). □

8.2 Analysis of the algebraic structure

The aim of this section is to establish a general control on $\partial_t\partial_t Z^I u_{\hat{\imath}}$, by making use of the identity (2.2.7). Our strategy is to express $\partial_t\partial_t Z^I u_{\hat{\imath}}$ by other terms arising in (2.2.7). In this first lemma, we relate $\partial_t\partial_t Z^I u_{\hat{\imath}}$ to remaining terms which have faster decay.

Lemma 8.2.1. *Let* $U_{tt}(I) := \big(\partial_t\partial_t Z^I u_1, \partial_t\partial_t Z^I u_2, \dots, \partial_t\partial_t Z^I u_{j_0}\big)^T$ *which is a* j_0*-vector and* $\mathbf{I}_{j_0}$ *be the* j_0*- dimensional identity matrix. The following identity holds:*

$$\begin{aligned}&\big((s/t)^2\mathbf{I}_{j_0} + \mathbf{G}(w,\partial w)\big)U_{tt}(I)\\&= \big([Z^I, G_{\hat{\imath}}^{\hat{\jmath}00}\partial_t\partial_t]u_{\hat{\jmath}} - Q_{G\hat{\imath}}(I,w,\partial w,\partial\partial w) + Z^I F_{\hat{\imath}} + R(Z^I u_{\hat{\imath}})\big)_{1\leqslant\hat{\imath}\leqslant j_0},\end{aligned} \tag{8.2.1}$$

where $\big(\mathbf{G}(w,\partial w)\big)_{1\leqslant\hat{\imath},\hat{\jmath}\leqslant j_0} = \big(\underline{G}_{\hat{\imath}}^{\hat{\jmath}00}(w,\partial w)\big)_{1\leqslant\hat{\imath},\hat{\jmath}\leqslant j_0}$ *is a* $j_0\times j_0$ *order matrix,*

$$\begin{aligned}&Q_{G\hat{\imath}}(I,w,\partial w,\partial\partial w)\\&= Z^I\big(\underline{G}_{\hat{\imath}}^{\hat{\jmath}a0}(w,\partial w)\underline{\partial}_a\partial_t u_{\hat{\jmath}} + \underline{G}_{\hat{\imath}}^{\hat{\jmath}0b}(w,\partial w)\partial_t\underline{\partial}_b u_{\hat{\jmath}} + \underline{G}_{\hat{\imath}}^{\hat{\jmath}ab}(w,\partial w)\underline{\partial}_a\underline{\partial}_b u_{\hat{\jmath}}\big)\\&\quad + Z^I\big(G_{\hat{\imath}}^{\hat{\jmath}\alpha\beta}(w,\partial w)\underline{\partial}_{\beta'}u_{\hat{\jmath}}\partial_\alpha\Psi_\beta^{\beta'}\big) + Z^I\big(G_{\hat{\imath}}^{\check{\jmath}\alpha\beta}(w,\partial w)\partial_\alpha\partial_\beta v_{\check{\jmath}}\big)\end{aligned}$$

and

$$\begin{aligned}R(Z^I u_{\hat{\imath}}) :=& -\underline{m}^{0a}\underline{\partial}_0\underline{\partial}_a Z^I u_{\hat{\imath}} - \underline{m}^{a0}\underline{\partial}_a\underline{\partial}_0 Z^I u_{\hat{\imath}} - \underline{m}^{ab}\underline{\partial}_a\partial_b Z^I u_{\hat{\imath}}\\&+ \underline{m}^{\alpha\beta}\big(\underline{\partial}_\alpha\Phi_\beta^{\beta'}\big)\underline{\partial}_{\beta'}Z^I u_{\hat{\imath}}.\end{aligned}$$

The terms $Z^I F_{\hat{\imath}}$ and $Q_{G\hat{\imath}}$ are bilinear and can be expected to enjoy better estimates. The term $R(Z^I u_{\hat{\imath}})$ contains only "good" second-order derivatives and can also be expected to enjoy better estimates.

Proof. Recall the identity (2.2.7) with the function $Z^I u_{\hat{\imath}}$

$$\begin{aligned}(s/t)^2\underline{\partial}_0\underline{\partial}_0 Z^I u_{\hat{\imath}} =& \Box Z^I u_{\hat{\imath}} - \underline{m}^{0a}\underline{\partial}_0\underline{\partial}_a Z^I u_{\hat{\imath}} - \underline{m}^{a0}\underline{\partial}_a\underline{\partial}_0 Z^I u_{\hat{\imath}}\\&- \underline{m}^{ab}\underline{\partial}_a\partial_b Z^I u_{\hat{\imath}} + \underline{m}^{\alpha\beta}\big(\underline{\partial}_\alpha\Phi_\beta^{\beta'}\big)\underline{\partial}_{\beta'}Z^I u_{\hat{\imath}}\\=:& \Box Z^I u_{\hat{\imath}} + R(Z^I, u_{\hat{\imath}}).\end{aligned} \tag{8.2.2}$$

By equation (1.2.1), the first term in the right-hand side can be written as

$$\Box Z^I u_{\hat{\imath}} = Z^I\Box u_{\hat{\imath}} = -Z^I\big(G_{\hat{\imath}}^{j\alpha\beta}(w,\partial w)\partial_\alpha\partial_\beta w_j\big) + Z^I(F_{\hat{\imath}}(w,\partial w)),$$

where

$$G_{\hat{\imath}}^{j\alpha\beta}(w,\partial w)\partial_\alpha\partial_\beta w_j = G_{\hat{\imath}}^{\hat{\jmath}\alpha\beta}(w,\partial w)\partial_\alpha\partial_\beta u_{\hat{\jmath}} + G_{\hat{\imath}}^{\check{\jmath}\alpha\beta}(w,\partial w)\partial_\alpha\partial_\beta v_{\check{\jmath}}$$

and

$$G_{\hat{\imath}}^{\hat{\jmath}\alpha\beta}(w,\partial w)\partial_\alpha\partial_\beta u_{\hat{\jmath}} = \underline{G}_{\hat{\imath}}^{\hat{\jmath}\alpha\beta}(w,\partial w)\underline{\partial}_\alpha\underline{\partial}_\beta u_{\hat{\jmath}} + G_{\hat{\imath}}^{\hat{\jmath}\alpha\beta}(w,\partial w)\underline{\partial}_{\beta'} u_{\hat{\jmath}}\partial_\alpha\Psi_\beta^{\beta'}.$$

We have

$$\begin{aligned}
&Z^I\big(G_{\hat{\imath}}^{\hat{\jmath}\alpha\beta}(w,\partial w)\partial_\alpha\partial_\beta u_{\hat{\jmath}}\big)\\
&= Z^I\big(\underline{G}_{\hat{\imath}}^{\hat{\jmath}\alpha\beta}(w,\partial w)\underline{\partial}_\alpha\underline{\partial}_\beta u_{\hat{\jmath}}\big) + Z^I\big(G_{\hat{\imath}}^{\hat{\jmath}\alpha\beta}(w,\partial w)\underline{\partial}_{\beta'} u_{\hat{\jmath}}\partial_\alpha\Psi_\beta^{\beta'}\big)\\
&= Z^I\big(\underline{G}_{\hat{\imath}}^{\hat{\jmath}00}\partial_t\partial_t u_{\hat{\jmath}} + \underline{G}_{\hat{\imath}}^{\hat{\jmath}a0}\underline{\partial}_a\partial_t u_{\hat{\jmath}} + \underline{G}_{\hat{\imath}}^{\hat{\jmath}0b}\partial_t\underline{\partial}_b u_{\hat{\jmath}} + \underline{G}_{\hat{\imath}}^{\hat{\jmath}ab}\underline{\partial}_a\underline{\partial}_b u_{\hat{\jmath}}\big)\\
&\quad + Z^I\big(G_{\hat{\imath}}^{\hat{\jmath}\alpha\beta}(w,\partial w)\underline{\partial}_{\beta'} u_{\hat{\jmath}}\partial_\alpha\Psi_\beta^{\beta'}\big)\\
&= \underline{G}_{\hat{\imath}}^{\hat{\jmath}00}(w,\partial w)\partial_t\partial_t Z^I u_{\hat{\jmath}} + [Z^I, \underline{G}_{\hat{\imath}}^{\hat{\jmath}00}(w,\partial w)\partial_t\partial_t]u_{\hat{\jmath}}\\
&\quad + Z^I\big(\underline{G}_{\hat{\imath}}^{\hat{\jmath}0b}(w,\partial w)\partial_t\underline{\partial}_b u_{\hat{\jmath}} + \underline{G}_{\hat{\imath}}^{\hat{\jmath}a0}(w,\partial w)\underline{\partial}_a\partial_t u_{\hat{\jmath}} + \underline{G}_{\hat{\imath}}^{\hat{\jmath}ab}(w,\partial w)\underline{\partial}_a\underline{\partial}_b u_{\hat{\jmath}}\big)\\
&\quad + Z^I\big(G_{\hat{\imath}}^{\hat{\jmath}\alpha\beta}(w,\partial w)\underline{\partial}_{\beta'} u_{\hat{\jmath}}\partial_\alpha\Psi_\beta^{\beta'}\big).
\end{aligned}$$

So we conclude that

$$\begin{aligned}
\Box Z^I u_{\hat{\imath}} = &- \underline{G}_{\hat{\imath}}^{\hat{\jmath}00}(w,\partial w)\partial_t\partial_t Z^I u_{\hat{\jmath}} - [Z^I, \underline{G}_{\hat{\imath}}^{\hat{\jmath}00}(w,\partial w)\partial_t\partial_t]u_{\hat{\jmath}}\\
&- Z^I\big(\underline{G}_{\hat{\imath}}^{\hat{\jmath}a0}(w,\partial w)\underline{\partial}_a\partial_t u_{\hat{\jmath}} + \underline{G}_{\hat{\imath}}^{\hat{\jmath}0b}(w,\partial w)\partial_t\underline{\partial}_b u_{\hat{\jmath}} + \underline{G}_{\hat{\imath}}^{\hat{\jmath}ab}(w,\partial w)\underline{\partial}_a\underline{\partial}_b u_{\hat{\jmath}}\big)\\
&- Z^I\big(G_{\hat{\imath}}^{\hat{\jmath}\alpha\beta}(w,\partial w)\underline{\partial}_{\beta'} u_{\hat{\jmath}}\partial_\alpha\Psi_\beta^{\beta'}\big) - Z^I\big(G_{\hat{\imath}}^{\check{\jmath}\alpha\beta}(w,\partial w)\partial_\alpha\partial_\beta v_{\check{\jmath}}\big)\\
&+ Z^I F_{\hat{\imath}}(w,\partial w).
\end{aligned}$$

Substituting this result into the equation (8.2.2), we obtain

$$\begin{aligned}
&(s/t)^2\partial_t\partial_t Z^I u_{\hat{\imath}} + \underline{G}_{\hat{\imath}}^{\hat{\jmath}00}(w,\partial w)\partial_t\partial_t Z^I u_{\hat{\jmath}}\\
&= -[Z^I, \underline{G}_{\hat{\imath}}^{\hat{\jmath}00}(w,\partial w)\partial_t\partial_t]u_{\hat{\jmath}}\\
&\quad - Z^I\big(\underline{G}_{\hat{\imath}}^{\hat{\jmath}a0}(w,\partial w)\underline{\partial}_a\partial_t u_{\hat{\jmath}} - \underline{G}_{\hat{\imath}}^{\hat{\jmath}0b}(w,\partial w)\partial_t\underline{\partial}_b u_{\hat{\jmath}} - \underline{G}_{\hat{\imath}}^{\hat{\jmath}ab}(w,\partial w)\underline{\partial}_a\underline{\partial}_b u_{\hat{\jmath}}\big)\\
&\quad - Z^I\big(G_{\hat{\imath}}^{\hat{\jmath}\alpha\beta}(w,\partial w)\underline{\partial}_{\beta'} u_{\hat{\jmath}}\partial_\alpha\Psi_\beta^{\beta'}\big) - Z^I\big(G_{\hat{\imath}}^{\check{\jmath}\alpha\beta}(w,\partial w)\partial_\alpha\partial_\beta v_{\check{\jmath}}\big)\\
&\quad + Z^I F_{\hat{\imath}}(w,\partial w) + R(Z^I, u_{\hat{\imath}}).
\end{aligned}$$

□

Now we derive estimates from the algebraic relation (8.2.1). A first step is to get the estimate of the inverse of the linear operator $(\mathbf{I}_{j_0} + (t/s)^2\mathbf{G})$. We can expect that when $|(t/s)^2\mathbf{G}|$ is small, $(\mathbf{I}_{j_0} + (t/s)^2\mathbf{G})$ is invertible and we can estimate $U_{tt}(I)$ from (8.2.1).

Lemma 8.2.2. *There exists a positive constant ϵ_0'', such that if the following sup-norm estimates*

$$\begin{aligned}
&|\partial u_{\hat{\imath}}| \leqslant CC_1\epsilon t^{-1/2}s^{-1}, \quad |\partial v_{\check{\imath}}| \leqslant CC_1\epsilon t^{-3/2}s^{\delta},\\
&|v_{\check{\imath}}| \leqslant CC_1\epsilon t^{-3/2}s^{\delta}, \quad |u_{\check{\imath}}| \leqslant CC_1\epsilon t^{-3/2}s
\end{aligned} \tag{8.2.3}$$

hold for $C_1\epsilon < \epsilon_0''$, *then the following estimate holds:*

$$
\begin{aligned}
\left|\partial_t\partial_t Z^I u_{\hat{\imath}}\right| \leqslant & C(t/s)^2 \max_{1\leqslant\hat{k}\leqslant j_0} \Big\{\Big(\left|[Z^I, G_{\hat{k}}^{\hat{\jmath}00}\partial_t\partial_t]u_{\hat{\jmath}}\right| + \left|Q_{G\hat{k}}(I, w, \partial w, \partial\partial w)\right| \\
& + \left|Z^I F_{\hat{k}}\right| + \left|R(Z^I u_{\hat{k}})\right|\Big)\Big\}.
\end{aligned}
\tag{8.2.4}
$$

Remark 8.2.1. (8.2.3) can be guaranteed by the energy assumption (2.4.5) via the L^∞ estimates (7.5.1), (7.5.2), (7.5.3), and (7.5.5) with C determined by the structure of the system.

Proof. By the structure of $\mathbf{G}(w, \partial w)$

$$
\underline{G}_{\hat{\imath}}^{\hat{\jmath}00} = \underline{A}_{\hat{\imath}}^{\hat{\jmath}00\gamma\hat{k}}\underline{\partial}_\gamma u_{\hat{k}} + \underline{A}_{\hat{\imath}}^{\hat{\jmath}00\gamma\check{k}}\underline{\partial}_\gamma v_{\check{k}} + \underline{B}_{\hat{\imath}}^{\hat{\jmath}00\hat{k}}u_{\hat{k}} + \underline{B}_{\hat{\imath}}^{\hat{\jmath}00\check{k}}v_{\check{k}}.
$$

Taking into account the assumption of (8.2.3),

$$
|\underline{G}_{\hat{\imath}}^{\hat{\jmath}00}| \leqslant CC_1\epsilon(s/t)^2.
$$

When $CC_1\epsilon \leqslant C\epsilon_0''$ sufficiently small, the linear operator $\mathbf{I}_{j_0} + (t/s)^2\mathbf{G}(w, \partial w)$ is invertible (viewed as a linear mapping from $(\mathbb{R}^{j_0}, \|\,\|_\infty)$ to itself) and $\|(\mathbf{I}_{j_0} + (t/s)^2\mathbf{G}(w, \partial w))^{-1}\|_{\infty,\infty}$ is bounded by a fixed constant.

By Lemma 8.2.1, we have

$$
\begin{aligned}
&(s/t)^2\partial_t\partial_t Z^I u_{\hat{\imath}} \\
&= (\mathbf{I}_{j_0} + (t/s)^2\mathbf{G}(w, \partial w))^{-1} \\
&\quad \Big([Z^I, G_{\hat{\imath}}^{\hat{\jmath}00}\partial_t\partial_t]u_{\hat{\jmath}} - Q_{G\hat{\imath}}(I, w, \partial w, \partial\partial w) + Z^I F_{\hat{\imath}} + R(Z^I u_{\hat{\imath}})\Big)_{1\leqslant\hat{\imath}\leqslant j_0}
\end{aligned}
$$

and so

$$
\begin{aligned}
\left|(s/t)^2\partial_t\partial_t Z^I u_{\hat{\imath}}\right| \leqslant & C \max_{1\leqslant\hat{k}\leqslant j_0} \Big(\left|[Z^I, G_{\hat{k}}^{\hat{\jmath}00}\partial_t\partial_t]u_{\hat{\jmath}}\right| + \left|Q_{G\hat{k}}(I, w, \partial w, \partial\partial w)\right|\Big) \\
& + C\Big(\left|Z^I F_{\hat{k}}\right| + \left|R(Z^I u_{,\hat{k}})\right|\Big).
\end{aligned}
$$

□

We need to estimate the terms appearing in the right-hand side of (8.2.4). We observe that the term $[Z^I, G_{\hat{\imath}}^{\hat{\jmath}00}\partial_t\partial_t]u_{\hat{\jmath}}$ contains the factor $\partial_t\partial_t Z^J u_{\hat{\jmath}}$ which is also a second-order derivative but with $|J| < |I|$. This structure leads us to the following induction estimates. Recall again our convention: $Z^I = 0$ when $|I| < 0$.

Lemma 8.2.3. *Followed by the notation of Lemma 8.2.1 and 8.2.2, the following induction estimate holds*

$$
\begin{aligned}
&|\partial_t\partial_t Z^I u_{\hat\imath}| \\
&\leqslant C(t/s)^2 \sum_{\substack{|I_1|+|I_2|\leqslant|I| \\ |I_1|<|I|,\gamma,\hat\jmath,i}} \big(|Z^{I_2}\partial_\gamma w_i| + |Z^{I_2} w_i|\big)|\partial_t\partial_t Z^{I_1} u_{\hat\jmath}| \\
&\quad + C(t/s)^2 \max_{1\leqslant\hat k\leqslant j_0} \big\{Q_{T\hat k}(I,w,\partial w,\partial\partial w) + C(t/s)^2|Q_{G\hat k}(I,w,\partial w\partial\partial w)| \\
&\quad + C(t/s)^2|Z^I F_{\hat k}| + C(t/s)^2|R(Z^I u_{\hat k})|\big\},
\end{aligned}
\tag{8.2.5}
$$

where

$$
Q_{T\hat\imath} = \sum_{\substack{|I_1|+|I_2|\leqslant|I| \\ |I_1|<|I|,\gamma,j,\hat\jmath}} \sum_{a,\beta,\gamma'} \big(|Z^{I_2}\partial_\gamma w_j| + |Z^{I_2} w_j|\big)\big(|\underline{\partial}_a\underline{\partial}_\beta Z^{I_1} u_{\hat\jmath}| + t^{-1}|\partial_{\gamma'} Z^{I_1} u_{\hat\jmath}|\big)
$$

Proof. This is purely an estimate for the term $[Z^I, \underline{G}_{\hat\imath}^{\hat\jmath 00}\partial_t\partial_t]u_{\hat\jmath}$, derived as follows:

$$
[Z^I, \underline{G}_{\hat\imath}^{\hat\jmath 00}\partial_t\partial_t]u_{\hat\jmath} = \sum_{\substack{I_1+I_2=I \\ |I_1|<|I|}} Z^{I_2}\big(\underline{G}_{\hat\imath}^{\hat\jmath 00}\big)\, Z^{I_1}\big(\partial_t\partial_t u_{\hat\jmath}\big) + \underline{G}_{\hat\imath}^{\hat\jmath 00}[Z^I, \partial_t\partial_t]u_{\hat\jmath}.
\tag{8.2.6}
$$

We have

$$
\begin{aligned}
\big|Z^{I_2}\big(\underline{G}_{\hat\imath}^{\hat\jmath 00}\big)\big| &\leqslant \big|Z^{I_2}\big(\underline{A}_{\hat\imath}^{\hat\jmath 00\gamma k}\underline{\partial}_\gamma w_k\big)\big| + \big|Z^{I_2}\big(\underline{B}_{\hat\imath}^{\hat\jmath 00k} w_k\big)\big| \\
&\leqslant C \sum_{\substack{j,\gamma \\ |I_3|\leqslant|I_2|}} \big(|Z^{I_3}\partial_\gamma w_j| + |Z^{I_3} w_j|\big).
\end{aligned}
$$

The first term of the right-hand side of (8.2.6) is bounded by

$$
\begin{aligned}
&C \sum_{\substack{|I_1|+|I_2|\leqslant|I| \\ |I_1|<|I|,\gamma,j,\hat\jmath}} \big(|Z^{I_2}\partial_\gamma w_j| + |Z^{I_2} w_j|\big)|\partial_t\partial_t Z^{I_1} u_{\hat\jmath}| \\
&+ C \sum_{\substack{|I_1|+|I_2|\leqslant|I| \\ |I_1|<|I|,\gamma,j,\hat\jmath}} \sum_{a,\beta,\gamma'} \big(|Z^{I_2}\partial_\gamma w_j| + |Z^{I_2} w_j|\big)\big(|\underline{\partial}_a\underline{\partial}_\beta Z^{I_1} u_{\hat\jmath}| + t^{-1}|\partial_{\gamma'} Z^{I_1} u_{\hat\jmath}|\big)
\end{aligned}
$$

Similarly, the second term of the right-hand side of (8.2.6) is estimated as follows, by (3.3.3) and Lemma 8.1.1:

$$
\begin{aligned}
\big|[Z^I, \partial_t\partial_t]u_{\hat\jmath}\big| &\leqslant C \sum_{\substack{\alpha,\beta \\ |J|<|I|}} |\partial_\alpha\partial_\beta Z^J u_{\hat\jmath}| \\
&\leqslant C \sum_{|J|<|I|} |\partial_t\partial_t Z^J u_{\hat\jmath}| + C \sum_{\substack{a,\beta \\ |J|<|I|}} |\underline{\partial}_a\underline{\partial}_\beta Z^J u_{\hat\jmath}| \\
&\quad + Ct^{-1} \sum_{\substack{\gamma \\ |J|<|I|}} |\partial_\gamma Z^J u_{\hat\jmath}|.
\end{aligned}
$$

Combined with the following estimate, we have

$$|\underline{G}_{\hat\imath}^{\hat\jmath 00}| \leqslant C \sum_{\hat\jmath,\gamma} \big(|\partial_\gamma w_j| + |w_j|\big).$$

□

8.3 Structure of the quadratic terms

The aim of this section is to analyze the structure of the quadratic terms $Q_{T\hat\imath}$, $Q_{G\hat\imath}$ and $Z^I F_{\hat\imath}$. We emphasize that some components of these terms satisfy the null condition and we can use it to make the estimates a bit simpler, but we prefer to *avoid the use of the null structure* here in order to show the independence of these analysis on this structure.

First, we note that the terms $Q_{T\hat\imath}$ are linear combinations of the following terms with constant coefficients with $|I_1| + |I_2| \leqslant |I|$ and $|I_1| < |I|$

$$\begin{aligned}
&|Z^{I_2}\partial_\gamma u_{\hat k}|\,|\underline{\partial}_a\underline{\partial}_\beta Z^{I_1} u_{\hat\jmath}|, \quad |Z^{I_2}\partial_\gamma v_{\check k}|\,|\underline{\partial}_a\underline{\partial}_\beta Z^{I_1} u_{\hat\jmath}|,\\
&|Z^{I_2} u_{\hat k}|\,|\underline{\partial}_a\underline{\partial}_\beta Z^{I_1} u_{\hat\jmath}|, \quad |Z^{I_2} v_{\check k}|\,|\underline{\partial}_a\underline{\partial}_\beta Z^{I_1} u_{\hat\jmath}|,\\
&t^{-1}|Z^{I_2}\partial_\gamma u_{\hat k}|\,|\partial_{\gamma'} Z^{I_1} u_{\hat\jmath}|, \quad t^{-1}|Z^{I_2}\partial_\gamma v_{\check k}|\,|\partial_{\gamma'} Z^{I_1} u_{\hat\jmath}|,\\
&t^{-1}|Z^{I_2} u_{\hat k}|\,|\partial_{\gamma'} Z^{I_1} u_{\hat\jmath}|, \quad t^{-1}|Z^{I_2} v_{\check k}|\,|\partial_{\gamma'} Z^{I_1} u_{\hat\jmath}|.
\end{aligned} \tag{8.3.1}$$

We consider $Q_{G\hat\imath}$ and write

$$\begin{aligned}
&Q_{G\hat\imath}(I, w, \partial w, \partial\partial w)\\
&= Z^I\big(\underline{G}_{\hat\imath}^{\hat\jmath a0}(w,\partial w)\underline{\partial}_a\partial_t u_{\hat\jmath} + \underline{G}_{\hat\imath}^{\hat\jmath 0b}(w,\partial w)\partial_t\underline{\partial}_b u_{\hat\jmath} + \underline{G}_{\hat\imath}^{\hat\jmath ab}(w,\partial w)\underline{\partial}_a\underline{\partial}_b u_{\hat\jmath}\big)\\
&\quad + Z^I\big(G_{\hat\imath}^{\hat\jmath\alpha\beta}(w,\partial w)\underline{\partial}_{\beta'} u_{\hat\jmath}\partial_\alpha \Psi_\beta^{\beta'}\big) + Z^I\big(G_{\hat\imath}^{\check\jmath\alpha\beta}(w,\partial w)\partial_\alpha\partial_\beta v_{\check\jmath}\big)\\
&= Z^I\big(\underline{A}_{\hat\imath}^{\hat\jmath a0\gamma\hat k}\underline{\partial}_\gamma u_{\hat k}\underline{\partial}_a\partial_t u_{\hat\jmath} + \underline{B}_{\hat\imath}^{\hat\jmath a0\hat k} u_{\hat k}\underline{\partial}_a\partial_t u_{\hat\jmath} + \underline{A}_{\hat\imath}^{\hat\jmath a0\gamma\check k}\underline{\partial}_\gamma v_{\check k}\underline{\partial}_a\partial_t u_{\hat\jmath}\\
&\qquad + \underline{B}_{\hat\imath}^{\hat\jmath a0\check k} v_{\check k}\underline{\partial}_a\partial_t u_{\hat\jmath}\big)\\
&\quad + Z^I\big(\underline{A}_{\hat\imath}^{\hat\jmath 0b\gamma\hat k}\underline{\partial}_\gamma u_{\hat k}\partial_t\underline{\partial}_b u_{\hat\jmath} + \underline{B}_{\hat\imath}^{\hat\jmath 0b\hat k} u_{\hat k}\partial_t\underline{\partial}_b u_{\hat\jmath} + \underline{A}_{\hat\imath}^{\hat\jmath 0b\gamma\check k}\underline{\partial}_\gamma v_{\check k}\partial_t\underline{\partial}_b u_{\hat\jmath}\\
&\quad + \underline{B}_{\hat\imath}^{\hat\jmath 0b\check k} v_{\check k}\partial_t\underline{\partial}_b u_{\hat\jmath}\big)\\
&\quad + Z^I\big(\underline{A}_{\hat\imath}^{\hat\jmath ab\gamma\hat k}\underline{\partial}_\gamma u_{\hat k}\underline{\partial}_a\underline{\partial}_b u_{\hat\jmath} + \underline{B}_{\hat\imath}^{\hat\jmath ab\hat k} u_{\hat k}\underline{\partial}_a\underline{\partial}_b u_{\hat\jmath} + \underline{A}_{\hat\imath}^{\hat\jmath ab\gamma\check k}\underline{\partial}_\gamma v_{\check k}\underline{\partial}_a\underline{\partial}_b u_{\hat\jmath}\\
&\quad + \underline{B}_{\hat\imath}^{\hat\jmath ab\check k} v_{\check k}\underline{\partial}_a\underline{\partial}_b u_{\hat\jmath}\big)\\
&\quad + Z^I\big(A_{\hat\imath}^{\hat\jmath\alpha\beta\gamma\hat k}\partial_\gamma u_{\hat k}\underline{\partial}_{\beta'} u_{\hat\jmath}\partial_\alpha\Psi_\beta^{\beta'} + B_{\hat\imath}^{\hat\jmath\alpha\beta\hat k} u_{\hat k}\underline{\partial}_{\beta'} u_{\hat\jmath}\partial_\alpha\Psi_\beta^{\beta'}\\
&\quad + A_{\hat\imath}^{\hat\jmath\alpha\beta\gamma\check k}\partial_\gamma v_{\check k}\underline{\partial}_{\beta'} u_{\hat\jmath}\partial_\alpha\Psi_\beta^{\beta'} + B_{\hat\imath}^{\hat\jmath\alpha\beta\check k} v_{\check k}\underline{\partial}_{\beta'} u_{\hat\jmath}\partial_\alpha\Psi_\beta^{\beta'}\big)\\
&\quad + Z^I\big(A_{\hat\imath}^{\check\jmath\alpha\beta\gamma\hat k}\partial_\gamma u_{\hat k}\partial_\alpha\partial_\beta v_{\check\jmath} + A_{\hat\imath}^{\check\jmath\alpha\beta\gamma\check k}\partial_\gamma v_{\check k}\partial_\alpha\partial_\beta v_{\check\jmath} + B_{\hat\imath}^{\check\jmath\alpha\beta\check k} v_{\check k}\partial_\alpha\partial_\beta v_{\check k}\big).
\end{aligned}$$

Let $K := \max_{\alpha,\beta,\gamma,i,j,k}\{A_i^{j\alpha\beta\gamma k}, B_i^{j\alpha\beta}\}$ and remark that by (2.2.2),

$$|Z^I \underline{A}_{\hat{\imath}}^{\hat{\jmath}\alpha\beta\gamma k}| + |Z^I \underline{B}_{\hat{\imath}}^{\hat{\jmath}\alpha\beta k}| \leqslant C(I)K.$$

We observe that the terms $Q_{G\hat{\imath}}(I, w, \partial w, \partial\partial w)$ are linear combinations with bounded coefficients of the following terms:

$$\begin{gathered}
Z^I\big(u_{\hat{\imath}}\underline{\partial}_a\underline{\partial}_\beta u_{\hat{\jmath}}\big),\quad Z^I\big(v_{\check{\imath}}\underline{\partial}_a\underline{\partial}_\beta u_{\hat{\jmath}}\big),\quad Z^I\big(u_{\hat{\imath}}\partial_t\underline{\partial}_a u_{\hat{\jmath}}\big),\quad Z^I\big(v_{\check{\imath}}\partial_t\underline{\partial}_a u_{\hat{\jmath}}\big),\\
Z^I\big(\underline{\partial}_\gamma u_{\hat{\imath}}\underline{\partial}_a\underline{\partial}_\beta u_{\hat{\jmath}}\big),\quad Z^I\big(\underline{\partial}_\gamma v_{\check{\imath}}\underline{\partial}_a\underline{\partial}_\beta u_{\hat{\jmath}}\big),\quad Z^I\big(\underline{\partial}_\gamma u_{\hat{\imath}}\partial_t\underline{\partial}_a u_{\hat{\jmath}}\big),\quad Z^I\big(\underline{\partial}_\gamma v_{\check{\imath}}\partial_t\underline{\partial}_a u_{\hat{\jmath}}\big),\\
Z^I\big(\partial_\alpha\Psi_\beta^{\beta'}u_{\hat{\imath}}\underline{\partial}_{\beta'}u_{\hat{\jmath}}\big),\quad Z^I\big(\partial_\alpha\Psi_\beta^{\beta'}v_{\check{\imath}}\underline{\partial}_{\beta'}u_{\hat{\jmath}}\big),\\
Z^I\big(\partial_\alpha\Psi_\beta^{\beta'}\partial_\gamma u_{\hat{\imath}}\underline{\partial}_{\beta'}u_{\hat{\jmath}}\big),\quad Z^I\big(\partial_\alpha\Psi_\beta^{\beta'}\partial_\gamma v_{\check{\imath}}\underline{\partial}_{\beta'}u_{\hat{\jmath}}\big),\\
Z^I\big(v_{\check{\imath}}\partial_\alpha\partial_\beta v_{\check{\jmath}}\big),\quad Z^I\big(\partial_\gamma u_{\hat{\imath}}\partial_\alpha\partial_\beta v_{\check{\jmath}}\big),\quad Z^I\big(\partial_\gamma v_{\check{\imath}}\partial_\alpha\partial_\beta v_{\check{\jmath}}\big).
\end{gathered} \tag{8.3.2}$$

For the term $Z^I F_{\hat{\imath}}$, recalling its definition

$$Z^I F_{\hat{\imath}} = Z^I\big(P_{\hat{\imath}}^{\alpha\beta jk}\partial_\alpha w_j\partial_\beta w_k + Q_{\hat{\imath}}^{\alpha j\check{k}}v_{\check{k}}\partial_\alpha w_j + R_{\hat{\imath}}^{\check{\jmath}\check{k}}v_{\check{\jmath}}v_{\check{k}}\big),$$

we classify its components into the following groups:

$$\partial_\alpha u_{\hat{\jmath}}\partial_\beta u_{\hat{k}},\quad \partial_\alpha v_{\check{\jmath}}\partial_\beta w_k,\quad v_{\check{k}}\partial_\alpha w_j,\quad v_{\check{\jmath}}v_{\check{k}}.$$

We regard $Z^I F_{\hat{\imath}}$ as a linear combination with constant coefficients bounded by K of the following terms:

$$\begin{gathered}
Z^I\big(\partial_\alpha u_{\hat{\jmath}}\partial_\beta u_{\hat{k}}\big),\quad Z^I\big(\partial_\alpha v_{\check{\jmath}}\partial_\beta w_k\big),\\
Z^I\big(v_{\check{k}}\partial_\alpha u_{\hat{\jmath}}\big),\quad Z^I\big(v_{\check{\jmath}}v_{\check{k}}\big).
\end{gathered} \tag{8.3.3}$$

The estimates on $\partial_t\partial_t Z^I u_{\hat{\imath}}$ turn out to be the estimates on these terms listed in (8.3.2) and (8.3.3) and $R(Z^I u_{\hat{k}})$.

8.4 L^∞ estimates

The purpose of the section is to establish the L^∞ estimates on $\partial_t\partial_t Z^J u_{\hat{\imath}}$ (and then $\partial_\alpha\partial_\beta Z^J u_{\hat{\imath}}$) under the energy assumption (2.4.5). We need to combine estimates on the terms $Q_{T\hat{\imath}}$, $Q_{G\hat{\imath}}$, $Z^{J^\dagger}F_{\hat{\imath}}$ and $R(Z^{J^\dagger}u_{\hat{\imath}})$ with Lemma 8.2.3. So, we first consider these terms in the following two lemmas.

Lemma 8.4.1. *Under the energy assumption* (2.4.5), *the following estimates hold for any* $|J^\dagger| \leqslant 2$

$$|Z^{J^\dagger}F_{\hat{\imath}}| \leqslant C(C_1\epsilon)^2 t^{-3/2}s^{-1+\delta}, \tag{8.4.1a}$$

$$|Q_{G\hat{\imath}}(J^\dagger, w, \partial w, \partial\partial w)| \leqslant C(C_1\epsilon)^2 t^{-3/2}s^{-1+\delta}, \tag{8.4.1b}$$

$$Q_{T\hat{\imath}}(J^\dagger, w, \partial w, \partial\partial w) \leqslant C(C_1\epsilon)^2 t^{-3/2}s^{-1+\delta}, \tag{8.4.1c}$$

$$|R(Z^{J^\dagger}u_{\hat{\imath}})| \leqslant CC_1\epsilon t^{-3/2}s^{-1+\delta}. \tag{8.4.1d}$$

Lemma 8.4.2. *Under the energy assumption* (2.4.5), *the following estimates hold for any* $|J| \leqslant 1$

$$|Z^J F_{\hat{\imath}}| \leqslant C(C_1\epsilon)^2 t^{-3/2} s^{-1+\delta/2}, \tag{8.4.2a}$$

$$|Q_{G\hat{\imath}}(J, w, \partial w, \partial\partial w)| \leqslant C(C_1\epsilon)^2 t^{-3/2} s^{-1+\delta/2}, \tag{8.4.2b}$$

$$Q_{T\hat{\imath}}(J, w, \partial w, \partial\partial w) \leqslant C(C_1\epsilon)^2 t^{-3/2} s^{-1+\delta/2}, \tag{8.4.2c}$$

$$|R(Z^J u_{\hat{\imath}})| \leqslant CC_1\epsilon t^{-3/2} s^{-1+\delta/2}. \tag{8.4.2d}$$

Lemma 8.4.3. *Under the energy assumption* (2.4.5), *the following estimates hold:*

$$|F_{\hat{\imath}}| \leqslant C(C_1\epsilon)^2 t^{-3/2} s^{-1}, \tag{8.4.3a}$$

$$|Q_{G\hat{\imath}}(0, w, \partial w, \partial\partial w)| \leqslant C(C_1\epsilon)^2 t^{-3/2} s^{-1}, \tag{8.4.3b}$$

$$Q_{T\hat{\imath}}(0, w, \partial w, \partial\partial w) \leqslant C(C_1\epsilon)^2 t^{-3/2} s, \tag{8.4.3c}$$

$$|R(u_{\hat{\imath}})| \leqslant CC_1\epsilon t^{-3/2} s^{-1}. \tag{8.4.3d}$$

The proofs of these three lemmas are essentially the same. We compute the relevant terms and express them as linear combinations of some bilinear terms, then suitably estimate each of them. The difference of the decay rate between these three lemmas is due to the difference of regularity.

Proof of Lemma 8.4.1. The control of $Q_{G\hat{\imath}}(J^\dagger, w, \partial w, \partial\partial w)$ is obtained as follows. Recall that $Q_{G\hat{\imath}}(J^\dagger, w, \partial w, \partial\partial w)$ is a linear combination of the terms listed in (8.3.2) with the index I replaced by $J^\dagger$. We estimate $Z^{J^\dagger}(u_{\hat{\imath}}\underline{\partial}_a\partial_t u_{\hat{\jmath}})$ in details:

$$\begin{aligned}
\left|Z^{J^\dagger}(u_{\hat{\imath}}\underline{\partial}_a\partial_t u_{\hat{\jmath}})\right| &\leqslant \sum_{J_2+J_3=J^\dagger} |Z^{J_2}u_{\hat{\imath}}||Z^{J_3}\underline{\partial}_a\partial_t u_{\hat{\jmath}}| \\
&= |u_{\hat{\imath}}||Z^{J^\dagger}\underline{\partial}_a\partial_t u_{\hat{\jmath}}| + |Z^{J_2}u_{\hat{\imath}}||Z^{J_3}\underline{\partial}_a\partial_t u_{\hat{\jmath}}| + |Z^{J^\dagger}u_{\hat{\imath}}||\underline{\partial}_a\partial_t u_{\hat{\jmath}}|.
\end{aligned}$$

When $J_2 = 0$ and $J_3 = J^\dagger$, we apply (7.5.5c) and (7.5.4a); when $J_3 = 0$ and $J_2 = J^\dagger$, we apply (7.5.5a) and (7.5.4c); and when $|J_2| = 1$ and $|J_3| = 1$, we apply (7.5.5c) and (7.5.4a);

$$\left|Z^{J^\dagger}(w_i\underline{\partial}_a\partial_t u_{\hat{\jmath}})\right| \leqslant C(C_1\epsilon)^2 t^{-3+\delta}.$$

For the other terms, we will specify the L^∞ estimates to be used but omit the details.

Here the in the different columns represent the different partitions of $J^\dagger = J_2 + J_3$. $(a, \leqslant b)$ means $|J_2| = a$, $|J_3| \leqslant b$. In the last column, we specify the decay rate obtained directly by applying the given inequalities (modulo the constant $C(C_1\epsilon)^2$). The first coefficient is the one we used for the first factor, while the second coefficient is used for the second factor:

Products	$(2, \leqslant 0)$	$(1, \leqslant 1)$	$(0, \leqslant 2)$	Decay rate
$u_{\hat\imath}\underline{\partial}_a\underline{\partial}_\beta u_{\hat\jmath}$	(7.5.5a), (7.5.4c)	(7.5.5c), (7.5.4a)	(7.5.5c), (7.5.4a)	$t^{-3}s^{\delta}$
$v_{\check\imath}\underline{\partial}_a\underline{\partial}_\beta u_{\hat\jmath}$	(7.5.1e), (7.5.4c)	(7.5.1e), (7.5.4a)	(7.5.1e), (7.5.4a)	$t^{-3}s^{-1+2\delta}$
$u_{\hat\imath}\partial_t\underline{\partial}_a u_{\hat\jmath}$	(7.5.5a), (7.5.4c)	(7.5.5c), (7.5.4a)	(7.5.5c), (7.5.4a)	$t^{-3}s^{\delta}$
$v_{\check\imath}\partial_t\underline{\partial}_a u_{\hat\jmath}$	(7.5.1e), (7.5.4c)	(7.5.1e), (7.5.4a)	(7.5.1e), (7.5.4a)	$t^{-3}s^{-1+2\delta}$
$\underline{\partial}_\gamma v_{\check\imath}\underline{\partial}_a\underline{\partial}_\beta u_{\hat\jmath}$	(7.5.2a), (7.5.4c)	(7.5.2a), (7.5.4a)	(7.5.2a), (7.5.4a)	$t^{-3}s^{-1+2\delta}$
$\underline{\partial}_\gamma u_{\hat\imath}\underline{\partial}_a\underline{\partial}_\beta u_{\hat\jmath}$	(7.5.1a), (7.5.4c)	(7.5.3a), (7.5.4a)	(7.5.3a), (7.5.4a)	$t^{-2}s^{-2+\delta}$

Products	$(2, \leqslant 0)$	$(1, \leqslant 1)$	$(0, \leqslant 2)$	Decay rate
$\underline{\partial}_\gamma u_{\hat\imath}\partial_t\underline{\partial}_a u_{\hat\jmath}$	(7.5.1a), (7.5.4c)	(7.5.3a), (7.5.4a)	(7.5.3a), (7.5.4a)	$t^{-2}s^{-2+\delta}$
$\underline{\partial}_\gamma v_{\check\imath}\partial_t\underline{\partial}_a u_{\hat\jmath}$	(7.5.2a), (7.5.4c)	(7.5.2a), (7.5.4a)	(7.5.2a), (7.5.4a)	$t^{-3}s^{-1+2\delta}$
$v_{\check\imath}\partial_\alpha\partial_\beta v_{\check\jmath}$	(7.5.1e), (7.5.2a)	(7.5.1e), (7.5.2a)	(7.5.1e), (7.5.2a)	$t^{-3}s^{2\delta}$
$\partial_\gamma u_{\hat\imath}\partial_\alpha\partial_\beta v_{\check\jmath}$	(7.5.1a), (7.5.2a)	(7.5.3a), (7.5.2a)	(7.5.3a), (7.5.2a)	$t^{-2}s^{-1+2\delta}$
$\partial_\gamma v_{\check\imath}\partial_\alpha\partial_\beta v_{\check\jmath}$	(7.5.2a), (7.5.2a)	(7.5.2a), (7.5.2a)	(7.5.2a), (7.5.2a)	$t^{-3}s^{2\delta}$

Taking into account the fact that $s \leqslant Ct \leqslant Cs^2$ and $\delta < 1/6$, we conclude that these terms are bounded by $C(C_1\epsilon)^2 t^{-3/2}s^{-1+\delta}$.

Again, we have the following four terms from $Q_{G\hat\imath}$, which are estimated separately:

$$Z^{J^\dagger}\big(\partial_\alpha\Psi_\beta^{\beta'} u_{\hat\imath}\underline{\partial}_{\beta'}u_{\hat\jmath}\big), \quad Z^{J^\dagger}\big(\partial_\alpha\Psi_\beta^{\beta'} v_{\check\imath}\underline{\partial}_{\beta'}u_{\hat\jmath}\big),$$
$$Z^{J^\dagger}\big(\partial_\alpha\Psi_\beta^{\beta'} \partial_\gamma u_{\hat\imath}\underline{\partial}_{\beta'}u_{\hat\jmath}\big), \quad Z^{J^\dagger}\big(\partial_\alpha\Psi_\beta^{\beta'} \partial_\gamma v_{\check\imath}\underline{\partial}_{\beta'}u_{\hat\jmath}\big).$$

By observing that $\big|Z^I\partial_\alpha\Psi_\beta^{\beta'}\big| \leqslant C(I)t^{-1}$, these terms can be estimated by $C(C_1\epsilon)^2t^{-3}s^{2\delta}$. We give the proof for $Z^{J_1}\big(\partial_\alpha\Psi_\beta^{\beta'} u_{\hat\imath}\underline{\partial}_{\beta'}u_{\hat\jmath}\big)$:

$$\begin{aligned}
&\big|Z^{J^\dagger}\big(\partial_\alpha\Psi_\beta^{\beta'} u_{\hat\imath}\underline{\partial}_{\beta'}u_{\hat\jmath}\big)\big| \leqslant \sum_{J_2+J_3+J_4=J_1} \big|Z^{J_4}\partial_\alpha\Psi_\beta^{\beta'}\big|\,\big|Z^{J_2}u_{\hat\imath}\big|\,\big|Z^{J_3}\underline{\partial}_{\beta'}u_{\hat\jmath}\big| \\
&\leqslant Ct^{-1}\sum_{|J_2|+|J_3|\leqslant|J^\dagger|} \big|Z^{J_2}u_{\hat\imath}\big|\,\big|Z^{J_3}\underline{\partial}_{\beta'}u_{\hat\jmath}\big| \leqslant Ct^{-1}C_1\epsilon t^{-3/2}s^{1+\delta}C_1\epsilon t^{-1/2}s^{-1+\delta} \\
&= C(C_1\epsilon)^2t^{-3}s^{2\delta} \leqslant C(C_1\epsilon)^2t^{-3/2}s^{-1+\delta},
\end{aligned}$$

where we recall that $\delta < 1/6$, and (7.5.5a) and (7.5.1a) are been used. The other terms are estimated similarly, and we omit the details and only list out the inequalities we use for each term and each partition of indices; cf. Table 1.

Table 1

Terms	$(2, \leqslant 0)$	$(1, \leqslant 1)$	$(0, \leqslant 2)$	Decay rate
$t^{-1} Z^{J_2} u_{\hat{\imath}} Z^{J_3} \underline{\partial}_{\beta'} u_{\hat{\jmath}}$	(7.5.5a), (7.5.3a)	(7.5.5c), (7.5.3a)	(7.5.5c), (7.5.1a)	$t^{-3} s^{\delta}$
$t^{-1} Z^{J_2} v_{\check{\imath}} Z^{J_3} \underline{\partial}_{\beta'} u_{\hat{\jmath}}$	(7.5.1e), (7.5.3a)	(7.5.1e), (7.5.3a)	(7.5.1e), (7.5.1a)	$t^{-3} s^{-1+2\delta}$
$t^{-1} Z^{J_2} \partial_\gamma u_{\hat{\imath}} Z^{J_3} \underline{\partial}_{\beta'} u_{\hat{\jmath}}$	(7.5.1a), (7.5.3a)	(7.5.3a), (7.5.3a)	(7.5.3a), (7.5.1a)	$t^{-2} s^{-2+\delta}$
$t^{-1} Z^{J_2} \partial_\gamma v_{\check{\imath}} Z^{J_3} \underline{\partial}_{\beta'} u_{\hat{\jmath}}$	(7.5.1a), (7.5.3a)	(7.5.1a), (7.5.3a)	(7.5.1a), (7.5.1a)	$t^{-2} s^{-2+\delta}$

Table 2

Products	$(2, \leqslant 0)$	$(1, \leqslant 1)$	$(0, \leqslant 2)$	Decay rate
$\underline{\partial}_a \underline{\partial}_\beta Z^{J_1} u_{\hat{\imath}} Z^{J_2} \partial_\gamma u_{\hat{k}}$	(7.4.5a), (7.5.3a)	(7.4.5a), (7.5.3a)	(7.4.5c), (7.5.1a)	$t^{-2} s^{-2+\delta}$
$\underline{\partial}_a \underline{\partial}_\beta Z^{J_1} u_{\hat{\imath}} Z^{J_2} \partial_\gamma v_{\check{k}}$	(7.4.5a), (7.5.2a)	(7.4.5a), (7.5.2a)	(7.4.5c), (7.5.2a)	$t^{-3} s^{-1+2\delta}$
$\underline{\partial}_a \underline{\partial}_\beta Z^{J_1} u_{\hat{\imath}} Z^{J_2} u_{\hat{k}}$	(7.4.5a), (7.5.5c)	(7.4.5a), (7.5.5c)	(7.4.5c), (7.5.5a)	$t^{-3} s^{\delta}$
$\underline{\partial}_a \underline{\partial}_\beta Z^{J_1} u_{\hat{\imath}} Z^{J_2} v_{\check{k}}$	(7.4.5a), (7.5.1e)	(7.4.5a), (7.5.1e)	(7.4.5c), (7.5.1e)	$t^{-3} s^{-1+2\delta}$

Table 3

Products	$(2, \leqslant 0)$	$(1, \leqslant 1)$	$(0, \leqslant 2)$	Decay rate
$t^{-1} \partial_{\gamma'} Z^{J_1} u_{\hat{\jmath}} Z^{J_2} \partial_\gamma u_{\hat{k}}$	(7.4.2a), (7.5.3a)	(7.4.4a), (7.5.3a)	(7.4.4a), (7.5.1a)	$t^{-2} s^{-2+\delta}$
$t^{-1} \partial_{\gamma'} Z^{J_1} u_{\hat{\jmath}} Z^{J_2} \partial_\gamma v_{\check{k}}$	(7.4.2a), (7.5.2a)	(7.4.4a), (7.5.2a)	(7.4.4a), (7.5.2a)	$t^{-3} s^{-1+2\delta}$
$t^{-1} \partial_{\gamma'} Z^{J_1} u_{\hat{\jmath}} Z^{J_2} u_{\hat{k}}$	(7.4.2a), (7.5.5c)	(7.4.4a), (7.5.5c)	(7.4.4a), (7.5.5a)	$t^{-3} s^{\delta}$
$t^{-1} \partial_{\gamma'} Z^{J_1} u_{\hat{\jmath}} Z^{J_2} v_{\check{k}}$	(7.4.2a), (7.5.1e)	(7.4.4a), (7.5.1e)	(7.4.4a), (7.5.1e)	$t^{-3} s^{-1+2\delta}$

The term $Q_{T\hat{\imath}}$ a linear combination of the terms presented in (8.3.1), with I_i replaced by J_i and $i = 1, 2$. Each term is estimated as for Q_G. We give the list of inequalities we use for every partition $|J_1| + |J_2| \leqslant |J^\dagger|$: in Table 2 and Table 3 the symbol $(a, \leqslant b)$ means $|J_1| = a, |J_2| \leqslant b$. We conclude with (8.4.1c).

The estimates on $Z^{J^\dagger} F_{\hat{\imath}}$ are essentially the same. Recall that $Z^{J^\dagger} F_{\hat{\imath}}$ is a linear combination of the terms listed in (8.3.3) with I replaced by $J^\dagger$. We write in details the estimate of the term $Z^{J^\dagger}\left(\partial_\alpha u_{\hat{\jmath}} \partial_\beta u_{\hat{k}}\right)$, as follows:

$$
\begin{aligned}
\left|Z^{J^\dagger}\left(\partial_\alpha u_{\hat{\jmath}} \partial_\beta u_{\hat{k}}\right)\right| \leqslant & |Z^{J^\dagger} \partial_\alpha u_{\hat{\jmath}}|\, |\partial_\beta u_{\hat{k}}| + |\partial_\alpha u_{\hat{\jmath}}|\, |Z^{J_1} \partial_\beta u_{\hat{k}}| \\
& + \sum_{\substack{|J_1|,|J_2| \leqslant 1 \\ J_2 + J_3 = J^\dagger}} |Z^{J_2} \partial_\alpha u_{\hat{\jmath}}|\, |Z^{J_3} \partial_\beta u_{\hat{k}}| \\
\leqslant & C C_1 \epsilon t^{-1/2} s^{-1+\delta}\, C C_1 \epsilon t^{-1/2} s^{-1} \\
& + C C_1 \epsilon t^{-1/2} s^{-1}\, C C_1 \epsilon t^{-1/2} s^{-1+\delta} \\
& + C C_1 \epsilon t^{-1/2} s^{-1}\, C C_1 \epsilon t^{-1/2} s^{-1} \\
\leqslant & C (C_1 \epsilon)^2 t^{-1} s^{-2+\delta} \\
\leqslant & C (C_1 \epsilon)^2 t^{-3/2} s^{-1+\delta}.
\end{aligned}
$$

For the three partitions of $J^\dagger = J_1 + J_2$, the L^∞ estimates we use are: when $|J_1| = 0$ and $|J_2| \leqslant |J^\dagger|$, we apply (7.5.3a) and (7.5.1a); when $|J_1| = |J_2| = 1$, we apply (7.5.3a) and (7.5.3a); when $|J_1| = |J^\dagger|$ and $|J_2| \leqslant 0$, we apply (7.5.1a) and (7.5.3a).

For the remaining terms, for each partition of J_1, we just list out the L^∞ estimates we use (with decay rate modulo a factor $C(C_1\epsilon)^2$):

Products	$(2, \leqslant 0)$	$(1, \leqslant 1)$	$(0, \leqslant 2)$	Decay rate
$\partial_\alpha u_{\hat{\jmath}} \partial_\beta u_{\hat{k}}$	(7.5.1a), (7.5.3a)	(7.5.3a), (7.5.3a)	(7.5.3a), (7.5.1a)	$t^{-1} s^{-2+\delta}$
$\partial_\alpha v_{\check{\jmath}} \partial_\beta w_k$	(7.5.2a), (7.5.1a)	(7.5.2a), (7.5.1a)	(7.5.2a), (7.5.1a)	$t^{-2} s^{-1+2\delta}$
$v_{\check{k}} \partial_\alpha w_j$	(7.5.1e), (7.5.1a)	(7.5.1e), (7.5.1a)	(7.5.1e), (7.5.1a)	$t^{-2} s^{-1+2\delta}$
$v_{\check{\jmath}} v_{\check{k}}$	(7.5.1e), (7.5.1e)	(7.5.1e), (7.5.1e)	(7.5.1e), (7.5.1e)	$t^{-3+2\delta}$

Combined with the condition $\delta < 1/6$ and the fact that $s \leqslant Ct \leqslant Cs^2$, the estimate (8.4.1a) is proved.

On the other hand, the estimate of $R(Z^{J_1} u_{\hat{\imath}})$ is a direct application of (7.4.5a). □

Proof of Lemma 8.4.2. The proof is essentially the same as that of (8.4.2), and the main difference lies in the inequalities we use for each term and partition of the index. We omit the details and present the inequalities we use, as follows.

For the estimates of $Q_{G\hat{\imath}}(J, w, \partial w, \partial\partial w)$, the inequalities we use are listed in:

Products	$(1, \leqslant 0)$	$(0, \leqslant 1)$	Decay rate
$u_{\hat{\imath}}\underline{\partial}_a\underline{\partial}_\beta u_{\hat{\jmath}}$	(7.5.5c), (7.5.4c)	(7.5.5c), (7.5.4b)	$t^{-3}s^{\delta/2}$
$v_{\check{\imath}}\underline{\partial}_a\underline{\partial}_\beta u_{\hat{\jmath}}$	(7.5.1f), (7.5.4c)	(7.5.1f), (7.5.4b)	$t^{-3}s^{-1+\delta}$
$u_{\hat{\imath}}\partial_t\underline{\partial}_a u_{\hat{\jmath}}$	(7.5.5c), (7.5.4c)	(7.5.5c), (7.5.4b)	$t^{-3}s^{\delta/2}$
$v_{\check{\imath}}\partial_t\underline{\partial}_a u_{\hat{\jmath}}$	(7.5.1f), (7.5.4c)	(7.5.1f), (7.5.4b)	$t^{-3}s^{-1+\delta}$
$\underline{\partial}_\gamma v_{\check{\imath}}\underline{\partial}_a\underline{\partial}_\beta u_{\hat{\jmath}}$	(7.5.2b), (7.5.4c)	(7.5.2b), (7.5.4b)	$t^{-3}s^{-1+\delta}$
$\underline{\partial}_\gamma u_{\hat{\imath}}\underline{\partial}_a\underline{\partial}_\beta u_{\hat{\jmath}}$	(7.5.3a), (7.5.4c)	(7.5.3a), (7.5.4b)	$t^{-2}s^{-2+\delta/2}$

Products	$(1, \leqslant 0)$	$(0, \leqslant 1)$	Decay rate
$\underline{\partial}_\gamma u_{\hat{\imath}}\partial_t\underline{\partial}_a u_{\hat{\jmath}}$	(7.5.3a), (7.5.4c)	(7.5.3a), (7.5.4b)	$t^{-2}s^{-2+\delta/2}$
$\underline{\partial}_\gamma v_{\check{\imath}}\partial_t\underline{\partial}_a u_{\hat{\jmath}}$	(7.5.2b), (7.5.4c)	(7.5.2b), (7.5.4b)	$t^{-3}s^{-1+\delta}$
$v_{\check{\imath}}\partial_\alpha\partial_\beta v_{\check{\jmath}}$	(7.5.1f), (7.5.2b)	(7.5.1f), (7.5.2b)	$t^{-3}s^{\delta}$
$\partial_\gamma u_{\hat{\imath}}\partial_\alpha\partial_\beta v_{\check{\jmath}}$	(7.5.3a), (7.5.2b)	(7.5.3a), (7.5.2b)	$t^{-2}s^{-1+\delta/2}$
$\partial_\gamma v_{\check{\imath}}\partial_\alpha\partial_\beta v_{\check{\jmath}}$	(7.5.2b), (7.5.2b)	(7.5.2b), (7.5.2b)	$t^{-3}s^{\delta}$

Terms	$(1, \leqslant 0)$	$(0, \leqslant 1)$	Decay rate
$t^{-1}Z^{J_2}u_{\hat{\imath}}Z^{J_3}\underline{\partial}_{\beta'}u_{\hat{\jmath}}$	(7.5.5c), (7.5.3a)	(7.5.5c), (7.5.3a)	t^{-3}
$t^{-1}Z^{J_2}v_{\check{\imath}}Z^{J_3}\underline{\partial}_{\beta'}u_{\hat{\jmath}}$	(7.5.1f), (7.5.3a)	(7.5.1f), (7.5.3a)	$t^{-3}s^{-1+\delta/2}$
$t^{-1}Z^{J_2}\partial_\gamma u_{\hat{\imath}}Z^{J_3}\underline{\partial}_{\beta'}u_{\hat{\jmath}}$	(7.5.3a), (7.5.3a)	(7.5.3a), (7.5.3a)	$t^{-2}s^{-2}$
$t^{-1}Z^{J_2}\partial_\gamma v_{\check{\imath}}Z^{J_3}\underline{\partial}_{\beta'}u_{\hat{\jmath}}$	(7.5.1b), (7.5.3a)	(7.5.1b), (7.5.3a)	$t^{-2}s^{-2+\delta/2}$

For the term $Q_{T\hat{\imath}}(J, w, \partial w, \partial\partial w)$, we have the list:

Products	$(1, \leqslant 0)$	$(0, \leqslant 1)$	Decay rate
$\underline{\partial}_a\underline{\partial}_\beta Z^{J_1}u_{\hat{\imath}}\, Z^{J_2}\partial_\gamma u_{\hat{k}}$	(7.4.5b), (7.5.3a)	(7.4.5c), (7.5.3a)	$t^{-2}s^{-2+\delta/2}$
$\underline{\partial}_a\underline{\partial}_\beta Z^{J_1}u_{\hat{\imath}}\, Z^{J_2}\partial_\gamma v_{\check{k}}$	(7.4.5b), (7.5.2b)	(7.4.5c), (7.5.2b)	$t^{-3}s^{-1+\delta}$
$\underline{\partial}_a\underline{\partial}_\beta Z^{J_1}u_{\hat{\imath}}\, Z^{J_2}u_{\hat{k}}$	(7.4.5b), (7.5.5c)	(7.4.5c), (7.5.5c)	$t^{-3}s^{\delta}$
$\underline{\partial}_a\underline{\partial}_\beta Z^{J_1}u_{\hat{\imath}}\, Z^{J_2}v_{\check{k}}$	(7.4.5b), (7.5.1f)	(7.4.5c), (7.5.1f)	$t^{-3}s^{-1+\delta}$

Terms	$(1, \leqslant 0)$	$(0, \leqslant 1)$	Decay rate
$t^{-1}\partial_{\gamma'}Z^{J_1}u_{\hat{\jmath}}\, Z^{J_2}\partial_\gamma u_{\hat{k}}$	(7.4.4a), (7.5.3a)	(7.4.4a), (7.5.3a)	$t^{-2}s^{-2}$
$t^{-1}\partial_{\gamma'}Z^{J_1}u_{\hat{\jmath}}\, Z^{J_2}\partial_\gamma v_{\check{k}}$	(7.4.4a), (7.5.2b)	(7.4.4a), (7.5.2b)	$t^{-3}s^{-1+\delta/2}$
$t^{-1}\partial_{\gamma'}Z^{J_1}u_{\hat{\jmath}}\, Z^{J_2}u_{\hat{k}}$	(7.4.4a), (7.5.5c)	(7.4.4a), (7.5.5c)	t^{-3}
$t^{-1}\partial_{\gamma'}Z^{J_1}u_{\hat{\jmath}}\, Z^{J_2}v_{\check{k}}$	(7.4.4a), (7.5.1f)	(7.4.4a), (7.5.1f)	$t^{-2}s^{-1+\delta/2}$

Finally, for the term $F_{\hat{\imath}}$ we have

Products	$(1, \leqslant 0)$	$(0, \leqslant 1)$	Decay rate
$\partial_\alpha u_{\hat{\jmath}}\partial_\beta u_{\hat{k}}$	(7.5.3a), (7.5.3a)	(7.5.3a), (7.5.3a)	$t^{-1}s^{-2}$
$\partial_\alpha v_{\check{\jmath}}\partial_\beta w_k$	(7.5.2b), (7.5.1b)	(7.5.2b), (7.5.1b)	$t^{-2}s^{-1+\delta}$
$v_{\check{k}}\partial_\alpha w_j$	(7.5.1f), (7.5.1b)	(7.5.1f), (7.5.1b)	$t^{-2}s^{-1+\delta}$
$v_{\check{\jmath}}v_{\check{k}}$	(7.5.1f), (7.5.1f)	(7.5.1f), (7.5.1f)	$t^{-3+\delta}$

On the other hand, the estimate of $R(Z^J u)$ is a direct result of (7.4.5b). □

Proof of Lemma 8.4.3. The proof is essentially the same as the proof of Lemma 8.4.1 but much easier. We analyze first $Q_{G\hat{\imath}}(w, \partial w, \partial\partial w)$. This is a direct application of the L^∞ estimates established earlier on the terms listed in (8.3.2) with $I = 0$. As done in the proof of Lemma 8.4.1, this gives the following list (with decay rate modulo $C(C_1\epsilon)^2$):

Terms	$(0,0)$	Decay rate
$u_{\hat{\imath}}\underline{\partial}_a\underline{\partial}_\beta u_{\hat{\jmath}}$	(7.5.5c), (7.5.4c)	t^{-3}
$v_{\check{\imath}}\underline{\partial}_a\underline{\partial}_\beta u_{\hat{\jmath}}$	(7.5.1e), (7.5.4c)	$t^{-3}s^{-1+\delta}$
$u_{\hat{\imath}}\partial_t\underline{\partial}_a u_{\hat{\jmath}}$	(7.5.5c), (7.5.4c)	t^{-3}
$v_{\check{\imath}}\partial_t\underline{\partial}_a u_{\hat{\jmath}}$	(7.5.1e), (7.5.4c)	$t^{-3}s^{-1+\delta}$
$\underline{\partial}_\gamma u_{\hat{\imath}}\underline{\partial}_a\underline{\partial}_\beta u_{\hat{\jmath}}$	(7.5.3a), (7.5.4c)	$t^{-2}s^{-2}$
$\underline{\partial}_\gamma v_{\check{\imath}}\underline{\partial}_a\underline{\partial}_\beta u_{\hat{\jmath}}$	(7.5.2a), (7.5.4c)	$t^{-3}s^{-1+\delta}$

Terms	$(0,0)$	Decay rate
$\underline{\partial}_\gamma u_{\hat{\imath}}\partial_t\underline{\partial}_a u_{\hat{\jmath}}$	(7.5.3a), (7.5.4c)	$t^{-2}s^{-2}$
$\underline{\partial}_\gamma v_{\check{\imath}}\partial_t\underline{\partial}_a u_{\hat{\jmath}}$	(7.5.2a), (7.5.4c)	$t^{-3}s^{-1+\delta}$
$v_{\check{\imath}}\partial_\alpha\partial_\beta v_{\check{\jmath}}$	(7.5.1e), (7.5.2a)	$t^{-3}s^{2\delta}$
$\partial_\gamma u_{\hat{\imath}}\partial_\alpha\partial_\beta v_{\check{\jmath}}$	(7.5.3a), (7.5.2a)	$t^{-2}s^{-1+\delta}$
$\partial_\gamma v_{\check{\imath}}\partial_\alpha\partial_\beta v_{\check{\jmath}}$	(7.5.2a), (7.5.2a)	$t^{-3}s^{2\delta}$

The following four terms

$$\partial_\alpha\Psi_\beta^{\beta'} u_{\hat{\imath}}\underline{\partial}_{\beta'}u_{\hat{\jmath}}, \quad \partial_\alpha\Psi_\beta^{\beta'} v_{\check{\imath}}\underline{\partial}_{\beta'}u_{\hat{\jmath}}, \quad \partial_\alpha\Psi_\beta^{\beta'}\partial_\gamma u_{\hat{\imath}}\underline{\partial}_{\beta'}u_{\hat{\jmath}}, \quad \partial_\alpha\Psi_\beta^{\beta'}\partial_\gamma v_{\check{\imath}}\underline{\partial}_{\beta'}u_{\hat{\jmath}}$$

are estimated by taking into account the additional decay supplied by the factor $|\Psi_\beta^{\beta'}| \leqslant Ct^{-1}$. The inequalities we use for each term are listed as follows:

Terms	$(0,0)$	Decay rate
$t^{-1}u_{\hat{\imath}}\underline{\partial}_{\beta'}u_{\hat{\jmath}}$	(7.5.5c), (7.5.3a)	t^{-3}
$t^{-1}v_{\check{\imath}}\underline{\partial}_{\beta'}u_{\hat{\jmath}}$	(7.5.1e), (7.5.3a)	$t^{-3}s^{-1+\delta}$
$t^{-1}\partial_\gamma u_{\hat{\imath}}\underline{\partial}_{\beta'}u_{\hat{\jmath}}$	(7.5.3a), (7.5.3a)	$t^{-2}s^{-2}$
$t^{-1}\partial_\gamma v_{\check{\imath}}\underline{\partial}_{\beta'}u_{\hat{\jmath}}$	(7.5.2a), (7.5.3a)	$t^{-3}s^{-1+\delta}$

The estimates of $Q_{T\check{\imath}}$ are similar. We establish the following list and omit the details:

Products	$(0,0)$	Decay rate
$\partial_\gamma u_{\hat{k}}\,\underline{\partial}_a\underline{\partial}_\beta u_{\hat{\imath}}$	(7.5.3a), (7.5.4c)	$t^{-2}s^{-2}$
$\partial_\gamma v_{\check{k}}\,\underline{\partial}_a\underline{\partial}_\beta u_{\hat{\imath}}$	(7.5.2a), (7.5.4c)	$t^{-3}s^{-1+\delta}$
$u_{\hat{k}}\,\underline{\partial}_a\underline{\partial}_\beta u_{\hat{\imath}}$	(7.5.5c), (7.5.4c)	t^{-3}
$v_{\check{k}}\,\underline{\partial}_a\underline{\partial}_\beta u_{\hat{\imath}}$	(7.5.1e), (7.5.4c)	$t^{-3}s^{-1+\delta}$

Products	(0,0)	Decay rate
$t^{-1}\partial_\gamma u_{\hat{k}}\,\partial_{\gamma'} u_{\hat{\jmath}}$	(7.5.3a), (7.5.3a)	$t^{-2}s^{-2}$
$t^{-1}\partial_\gamma v_{\check{k}}\,\partial_{\gamma'} u_{\hat{\jmath}}$	(7.5.2a), (7.5.3a)	$t^{-3}s^{-1+\delta}$
$t^{-1}u_{\hat{k}}\,\partial_{\gamma'} u_{\hat{\jmath}}$	(7.5.5c), (7.5.3a)	t^{-3}
$t^{-1}v_{\check{k}}\,\partial_{\gamma'} u_{\hat{\jmath}}$	(7.5.1e), (7.5.3a)	$t^{-3}s^{-1+\delta}$

We conclude with (8.4.3c).

The estimate of $F_{\hat{\imath}}$ is as follows: recall the structure of $F_{\hat{\imath}}$ described by (8.3.3), we need to estimate these terms with $I = 0$. As in the proof of Lemma 8.4.1, the following list is established (with decay rate modulo $C(C_1\epsilon)^2$):

Products	(0,0)	Decay rate
$\partial_\alpha u_{\hat{\imath}}\partial_\beta u_{\hat{\jmath}}$	(7.5.3a), (7.5.3a)	$t^{-1}s^{-2}$
$\partial_\alpha v_{\check{\imath}}\partial_\beta w_j$	(7.5.2a), (7.5.1a)	$t^{-2}s^{-1+2\delta}$
$v_{\check{\imath}}\partial_\alpha w_j$	(7.5.1e), (7.5.1a)	$t^{-2}s^{-1+2\delta}$
$v_{\check{\imath}}v_{\check{\jmath}}$	(7.5.1e), (7.5.1e)	$t^{-3}s^{2\delta}$

By taking into account the condition $\delta < 1/6$ and the fact that $C^{-1}s \leqslant t \leqslant Cs^2$, we conclude with (8.4.3a).

The estimate of $R(u_{\hat{\imath}})$ is a direct application of (7.5.4c). □

We can now prove the main result of this section.

Proposition 8.4.1. *Let w_i the solution of* (1.2.1) *and assume that* (2.4.5) *holds with $C_1\epsilon \leqslant \min\{1, \epsilon_0''\}$. For the wave components, the following decay estimates hold for any $J^\dagger \leqslant 2$ and $|J| \leqslant 1$:*

$$\sup_{\mathcal{H}_s} \left(s^3 t^{-1/2} |\partial_t\partial_t Z^{J^\dagger} u_{\hat{\imath}}|\right) \leqslant CC_1\epsilon s^\delta, \tag{8.4.4a}$$

$$\sup_{\mathcal{H}_s} \left(s^3 t^{-1/2} |\partial_t\partial_t Z^{J} u_{\hat{\imath}}|\right) \leqslant CC_1\epsilon s^{\delta/2}, \tag{8.4.4b}$$

$$\sup_{\mathcal{H}_s} \left(s^3 t^{-1/2} |\partial_t\partial_t u_{\hat{\imath}}|\right) \leqslant CC_1\epsilon, \tag{8.4.4c}$$

and more generally

$$\sup_{\mathcal{H}_s} \left(s^3 t^{-1/2} |\partial_\alpha\partial_\beta Z^{J^\dagger} u|\right) \leqslant CC_1\epsilon s^\delta, \tag{8.4.5a}$$

$$\sup_{\mathcal{H}_s} \left(s^3 t^{-1/2} |\partial_\alpha\partial_\beta Z^{J} u|\right) \leqslant CC_1\epsilon s^{\delta/2}, \tag{8.4.5b}$$

$$\sup_{\mathcal{H}_s} \left(s^3 t^{-1/2} |\partial_\alpha\partial_\beta u|\right) \leqslant CC_1\epsilon. \tag{8.4.5c}$$

Proof. (8.4.4c) will be proved first. Substitute (8.4.3) into (8.2.5) with $|I| = 0$. Recall the convention $|I| < 0 \Rightarrow Z^I = 0$. The desired result is proven.

Observe that for (8.4.4b), the case $|J| \leqslant 0$ is proved by (8.4.4c). For the case $|J| = 1$, we recall (8.2.5):

$$\begin{aligned}|\partial_t\partial_t Z^J u_{\hat{\imath}}| \leqslant& C(t/s)^2 \sum_{\substack{|J_2|=1\\ \gamma,\hat{\jmath},i}} \big(|Z^{J_2}\partial_\gamma w_i| + |Z^{J_2} w_i|\big)|\partial_t\partial_t u_{\hat{\jmath}}| \\ &+ C(t/s)^2 Q_{T\hat{\imath}}(J_2, w, \partial w, \partial\partial w) + C(t/s)^2 |Q_{G\hat{\imath}}(J_2, w, \partial w\partial\partial w)| \\ &+ C(t/s)^2 |Z^{J_2} F_{\hat{\imath}}| + C(t/s)^2 |R(Z^{J_2} u_{\hat{\imath}})|\end{aligned}$$

substitutes (8.4.4c) (an estimate on $\partial_t\partial_t u_{\hat{\jmath}}$) and (8.4.2) into (8.2.5), together with the following estimate:

$$\sum_{\substack{\gamma\,\hat{\jmath},\hat{\imath}\\ |J|\leqslant 1}} \big(|Z^J\partial_\gamma w_i| + |Z^J w_i|\big) \leqslant CC_1\epsilon\big(t^{-1/2}s^{-1} + t^{-3/2}s + t^{-3/2}s^{\delta/2}\big) \leqslant CC_1\epsilon t^{-3/2}s.$$

For (8.4.4a), remark that the case $|J^\dagger| \leqslant 1$ is guaranteed by (8.4.4b) and (8.4.4c).

The case $|J^\dagger| = 2$ is done by substituting (8.4.4b) (with $|J| = 1$), (8.4.4c) and (8.4.1) into (8.2.5) with $|I| = 2$ together with the following estimate which is a direct result of (7.5.1a), (7.5.5c) and (7.5.1e):

$$\begin{aligned}\sum_{\gamma\,\hat{\jmath},\hat{\imath}} \big(|Z^{J^\dagger}\partial_\gamma w_i| + |Z^{J^\dagger} w_i|\big) &\leqslant CC_1\epsilon\big(t^{-1/2}s^{-1+\delta/2} + t^{-3/2}s + t^{-3/2}s^{\delta/2}\big) \\ &\leqslant CC_1\epsilon t^{-3/2}s^{1+\delta/2}.\end{aligned}$$

We recall (8.2.5) with $I = J^\dagger$:

$$\begin{aligned}&|\partial_t\partial_t Z^{J^\dagger} u_{\hat{\imath}}| \\ \leqslant& C(t/s)^2 \sum_{\substack{|J_1|+|J_2|\leqslant 2\\ |J_1|\leqslant 1,\gamma,\hat{\jmath},i}} \big(|Z^{J_2}\partial_\gamma w_i| + |Z^{J_2} w_i|\big)\, |\partial_t\partial_t Z^{J_1} u_{\hat{\jmath}}| \\ &+ C(t/s)^2 Q_{T\hat{\imath}}(J_2, w, \partial w, \partial\partial w) + C(t/s)^2 |Q_{G\hat{\imath}}(J_2, w, \partial w\partial\partial w)| \\ &+ C(t/s)^2 |Z^{J_2} F_{\hat{\imath}}| + C(t/s)^2 |R(Z^{J_2} u_{\hat{\imath}})|.\end{aligned}$$

Substituting (8.4.4b) (with $|J|=1$), (8.4.4c) and (8.4.1), we find that

$$
\begin{aligned}
&|\partial_t\partial_t Z^{J^\dagger} u_{\hat{\imath}}| \\
&\leqslant C(t/s)^2 \sum_{\substack{|J_1|+|J_2|\leqslant 2\\ |J_1|\leqslant 1,\gamma,\hat{\jmath},i}} \big(|Z^{J_2}\partial_\gamma w_i| + |Z^{J_2} w_i|\big)\,|\partial_t\partial_t Z^{J_1} u_{\hat{\jmath}}| + CC_1\epsilon t^{1/2} s^{-3+\delta} \\
&\leqslant C(t/s)^2 \sum_{\substack{|J_1|+|J_2|\leqslant 2\\ |J_1|=1,\gamma,\hat{\jmath},i}} \big(|Z^{J_2}\partial_\gamma w_i| + |Z^{J_2} w_i|\big)\,|\partial_t\partial_t Z^{J_1} u_{\hat{\jmath}}| \\
&\quad + C(t/s)^2 \sum_{\substack{|J_1|+|J_2|\leqslant 2\\ |J_1|=0,\gamma,\hat{\jmath},i}} \big(|Z^{J_2}\partial_\gamma w_i| + |Z^{J_2} w_i|\big)\,|\partial_t\partial_t Z^{J_1} u_{\hat{\jmath}}| + CC_1\epsilon 2 t^{1/2} s^{-3+\delta} \\
&\leqslant CC_1\epsilon (t/s)^2 t^{-3/2} s\, t^{1/2} s^{-3+\delta/2} + CC_1\epsilon (t/s)^2 t^{-3/2} s^{1+\delta/2}\, t^{1/2} s^{-3} \\
&\quad + CC_1\epsilon t^{1/2} s^{-3+\delta} \\
&\leqslant CC_1\epsilon t s^{-4+\delta/2} + CC_1\epsilon t^{1/2} s^{-3+\delta} \leqslant CC_1\epsilon t^{1/2} s^{-3+\delta}.
\end{aligned}
$$

The bound (8.4.5) are direct result of (8.4.4) combined with (8.1.2). □

We can give the complete L^∞ estimates of the second-order derivatives.

Proposition 8.4.2. *By relying on* (2.4.5) *with* $C_1\epsilon \leqslant \min\{1, \epsilon_0''\}$, *the following estimates hold for all* $|J^\dagger| \leqslant 2$ *and* $|J| \leqslant 1$*:*

$$\sup_{\mathcal{H}_s} |s^3 t^{-1/2} \partial_\alpha\partial_\beta Z^{J^\dagger} u| + \sup_{\mathcal{H}_s} |s^3 t^{-1/2} Z^{J^\dagger} \partial_\alpha\partial_\beta u| \leqslant CC_1\epsilon s^\delta, \tag{8.4.6a}$$

$$\sup_{\mathcal{H}_s} |s^3 t^{-1/2} \partial_\alpha\partial_\beta Z^{J} u| + \sup_{\mathcal{H}_s} |s^3 t^{-1/2} Z^{J} \partial_\alpha\partial_\beta u| \leqslant CC_1\epsilon s^{\delta/2}, \tag{8.4.6b}$$

$$\sup_{\mathcal{H}_s} |s^3 t^{-1/2} \partial_\alpha\partial_\beta u| \leqslant CC_1\epsilon. \tag{8.4.6c}$$

Proof. These estimates are a consequence of (3.3.3) and (8.4.5). □

8.5 L^2 estimates

The aim of this section is to get the L^2 estimates on $\partial_t\partial_t Z^I u_{\hat{\jmath}}$. As in the last section, the strategy is to make use of (8.2.5). First, we estimate the terms $Q_{T\hat{\imath}}$, $Q_{G\hat{\imath}}$, $Z^I F_{\hat{\imath}}$ and $R(Z^I u_{\hat{\imath}})$.

Lemma 8.5.1. *Under the energy assumption* (2.4.5), *the following estimates hold for all* $|I^\dagger| \leqslant 4$*:*

$$\big\|s Q_{G\hat{\imath}}(I^\dagger, w, \partial w, \partial\partial w)\big\|_{L^2(\mathcal{H}_s)} \leqslant C(C_1\epsilon)^2 s^\delta, \tag{8.5.1a}$$

$$\big\|sQ_{T\hat{\imath}}(I^\dagger, w, \partial w, \partial\partial w)\big\|_{L^2(\mathcal{H}_s)} \leqslant C(C_1\epsilon)^2 s^\delta, \tag{8.5.1b}$$

$$\big\|sZ^{I^\dagger} F_{\hat{\imath}}\big\|_{L^2(\mathcal{H}_s)} \leqslant C(C_1\epsilon)^2 s^\delta, \tag{8.5.1c}$$

$$\big\|sR(Z^{I^\dagger} u_{\hat{\imath}})\big\|_{L^2(\mathcal{H}_s)} \leqslant CC_1\epsilon s^\delta. \tag{8.5.1d}$$

Lemma 8.5.2. *Under the energy assumption* (2.4.5), *the following estimates hold for all* $|I| \leqslant 3$*:*

$$\big\|sQ_{G\hat{\imath}}(I, w, \partial w, \partial\partial w)\big\|_{L^2(\mathcal{H}_s)} \leqslant C(C_1\epsilon)^2 s^{\delta/2}, \tag{8.5.2a}$$

$$\big\|sQ_{T\hat{\imath}}(I, w, \partial w, \partial\partial w)\big\|_{L^2(\mathcal{H}_s)} \leqslant C(C_1\epsilon)^2 s^{\delta/2}, \tag{8.5.2b}$$

$$\big\|sZ^{I} F_{\hat{\imath}}\big\|_{L^2(\mathcal{H}_s)} \leqslant C(C_1\epsilon)^2 s^{\delta/2}, \tag{8.5.2c}$$

$$\big\|sR(Z^{I} u_{\hat{\imath}})\big\|_{L^2(\mathcal{H}_s)} \leqslant CC_1\epsilon s^{\delta/2}. \tag{8.5.2d}$$

Lemma 8.5.3. *Under the energy assumption* (2.4.5), *the following estimates hold for all* $|I^\flat| \leqslant 2$*:*

$$\big\|sQ_{G\hat{\imath}}(I^\flat, w, \partial w, \partial\partial w)\big\|_{L^2(\mathcal{H}_s)} \leqslant C(C_1\epsilon)^2, \tag{8.5.3a}$$

$$\big\|sQ_{G\hat{\imath}}(I^\flat, w, \partial w, \partial\partial w)\big\|_{L^2(\mathcal{H}_s)} \leqslant C(C_1\epsilon)^2, \tag{8.5.3b}$$

$$\big\|sZ^{I^\flat} F_{\hat{\imath}}\big\|_{L^2(\mathcal{H}_s)} \leqslant C(C_1\epsilon)^2, \tag{8.5.3c}$$

$$\big\|sR(Z^{I^\flat} u_{\hat{\imath}})\big\|_{L^2(\mathcal{H}_s)} \leqslant CC_1\epsilon. \tag{8.5.3d}$$

Proof of Lemma 8.5.1. We consider first $Q_{G\hat{\imath}}$ and recall the structure of $Q_{G\hat{\imath}}$ expressed by (8.3.2). We do an L^2 estimate on each term of (8.3.2) with $I = I^\dagger, |I^\dagger| \leqslant 4$. We take $u_{\hat{\imath}}\underline{\partial}_a\partial_t u_{\hat{\jmath}}$ as an example and we write down of the argument:

$$\begin{aligned}
&\big\|sZ^{I^\dagger}\big(u_{\hat{\imath}}\underline{\partial}_a\partial_t u_{\hat{\jmath}}\big)\big\|_{L^2(\mathcal{H}_s)}\\
&\leqslant \sum_{I_1+I_2=I^\dagger} \big\|s\big(Z^{I_1}u_{\hat{\imath}}\, Z^{I_2}\underline{\partial}_a\partial_t u_{\hat{\jmath}}\big)\big\|_{L^2(\mathcal{H}_s)}\\
&\leqslant \big\|s\big(u_{\hat{\imath}}\, Z^{I^\dagger}\underline{\partial}_a\partial_t u_{\hat{\jmath}}\big)\big\|_{L^2(\mathcal{H}_s)} + \sum_{\substack{|I_1|=1\\ I_1+I_2=I^\dagger}} \big\|s\big(Z^{I_1}u_{\hat{\imath}}\, Z^{I_2}\underline{\partial}_a\partial_t u_{\hat{\jmath}}\big)\big\|_{L^2(\mathcal{H}_s)}\\
&\quad + \sum_{\substack{|I_1|=2\\ I_1+I_2=I^\dagger}} \big\|s\big(Z^{I_1}u_{\hat{\imath}}\, Z^{I_2}\underline{\partial}_a\partial_t u_{\hat{\jmath}}\big)\big\|_{L^2(\mathcal{H}_s)} + \sum_{\substack{|I_1|=3\\ I_1+I_2=I^\dagger}} \big\|s\big(Z^{I_1}u_{\hat{\imath}}\, Z^{I_2}\underline{\partial}_a\partial_t u_{\hat{\jmath}}\big)\big\|_{L^2(\mathcal{H}_s)}\\
&\quad + \big\|s\big(Z^{I^\dagger}u_{\hat{\imath}}\,\underline{\partial}_a\partial_t u_{\hat{\jmath}}\big)\big\|_{L^2(\mathcal{H}_s)}\\
&=: T_0 + T_1 + T_2 + T_3 + T_4.
\end{aligned}$$

Observe now that in term T_k, $|I_2| \leqslant 4-k$.

The term T_0 is estimated by (7.5.5c) and (7.3.5a):

$$\begin{aligned}T_0 =&\|s(u_{\hat\imath}\, Z^{I_1}\underline{\partial}_a\partial_t u_{\hat\jmath})\|_{L^2(\mathcal{H}_s)} \leqslant CC_1\epsilon\|s(t^{-3/2}\, sZ^{I_1}\underline{\partial}_a\partial_t u_{\hat\jmath})\|_{L^2(\mathcal{H}_s)}\\ \leqslant& CC_1\epsilon s^{-1/2}\|sZ^{I_1}\underline{\partial}_a\partial_t u_{\hat\jmath}\|_{L^2(\mathcal{H}_s)}\\ \leqslant& C(C_1\epsilon)^2 s^{-1/2+\delta}.\end{aligned}$$

The term T_1 is estimated by (7.5.5c) and (7.3.5a):

$$\begin{aligned}T_1 =& \sum_{\substack{|I_1|=1\\ I_1+I_2=I^\dagger}} \|s(Z^{I_1}u_{\hat\imath}\, Z^{I_2}\underline{\partial}_a\partial_t u_{\hat\jmath})\|_{L^2(\mathcal{H}_s)}\\ \leqslant& CC_1\epsilon \sum_{\substack{|I_1|=1\\ I_1+I_2=I^\dagger}} \|s(t^{-3/2}\, sZ^{I_2}\underline{\partial}_a\partial_t u_{\hat\jmath})\|_{L^2(\mathcal{H}_s)}\\ \leqslant& CC_1\epsilon s^{-1/2} \sum_{\substack{|I_1|=1\\ I_1+I_2=I^\dagger}} \|sZ^{I_3}\underline{\partial}_a\partial_t u_{\hat\jmath}\|_{L^2(\mathcal{H}_s)} \leqslant C(C_1\epsilon)^2 s^{-1/2+\delta}.\end{aligned}$$

The term T_2 is estimated by (7.5.5a) and (7.3.5c) (remark that $|I_2| \leqslant 4-2=2$):

$$\begin{aligned}T_2 =& \sum_{\substack{|I_1|=2\\ I_1+I_2=I^\dagger}} \|s(Z^{I_1}u_{\hat\imath}\, Z^{I_2}\underline{\partial}_a\partial_t u_{\hat\jmath})\|_{L^2(\mathcal{H}_s)}\\ \leqslant& CC_1\epsilon \sum_{\substack{|I_1|=2\\ I_1+I_2=I^\dagger}} \|s(t^{-3/2}s^{1+\delta}\, Z^{I_2}\underline{\partial}_a\partial_t u_{\hat\jmath})\|_{L^2(\mathcal{H}_s)}\\ =& CC_1\epsilon s^{-1/2+\delta} \sum_{\substack{|I_1|=2\\ I_1+I_2=I^\dagger}} \|sZ^{I_2}\underline{\partial}_a\partial_t u_{\hat\jmath}\|_{L^2(\mathcal{H}_s)} \leqslant C(C_1\epsilon)^2 s^{-1/2+\delta}.\end{aligned}$$

The term T_3 is estimated by (7.5.5a) and (7.3.5c) (remark that $|I_2| \leqslant 4-3=1$):

$$\begin{aligned}T_3 =& \sum_{\substack{|I_1|=3\\ I_1+I_2=I^\dagger}} \|s(Z^{I_1}u_{\hat\imath}\, Z^{I_2}\underline{\partial}_a\partial_t u_{\hat\jmath})\|_{L^2(\mathcal{H}_s)}\\ \leqslant& CC_1\epsilon \sum_{\substack{|I_1|=3\\ I_1+I_2=I^\dagger}} \|s(t^{-3/2}s^{\delta}\, sZ^{I_2}\underline{\partial}_a\partial_t u_{\hat\jmath})\|_{L^2(\mathcal{H}_s)}\\ =& CC_1\epsilon s^{-1/2+\delta} \sum_{\substack{|I_1|=3\\ I_1+I_2=I^\dagger}} \|sZ^{I_2}\underline{\partial}_a\partial_t u_{\hat\jmath}\|_{L^2(\mathcal{H}_s)} \leqslant C(C_1\epsilon)^2 s^{-1/2+\delta}.\end{aligned}$$

The term T_4 is estimated by (7.3.6a) and (7.5.4c):

$$\begin{aligned}T_4 =& \|s(Z^{I^\dagger}u_{\hat\imath}\,\underline{\partial}_a\partial_t u_{\hat\jmath})\|_{L^2(\mathcal{H}_s)} \leqslant CC_1\epsilon\|sZ^{I^\dagger}u_{\hat\imath}\, t^{-3/2}s^{-1}\|_{L^2(\mathcal{H}_s)}\\ =& CC_1\epsilon s^{-1/2}\|t^{-1}Z^{I^\dagger}u_{\hat\imath}\|_{L^2(\mathcal{H}_s)} \leqslant CC_1\epsilon s^{-1/2}\|s^{-1}Z^{I^\dagger}u_{\hat\imath}\|_{L^2(\mathcal{H}_s)}\\ \leqslant& C(C_1\epsilon)^2 s^{-1/2+\delta}.\end{aligned}$$

So we conclude with

$$\left\| sZ^{I^\dagger}\left(u_{\hat\imath}\underline{\partial}_a\partial_t u_{\hat\jmath}\right)\right\|_{L^2(\mathcal{H}_s)} \leqslant C(C_1\epsilon)^2 s^{-1/2+\delta}.$$

For the remaining terms, we will not write the details, but for each partition of $I^\dagger = I_1 + I_2$ we give the L^2 and L^∞ estimates in Chapter 7. As in the estimate of $u_{\hat\imath}\underline{\partial}_a\partial_t u_{\hat\jmath}$, we denote by $(k, \leqslant 4-k)$ the terms with which $|I_1| = k$, $|I_2| \leqslant 4-k$. This leads us to Table 4 and Table 5.

There are four terms to be estimated separately:

$$sZ^{I^\dagger}\left(\partial_\alpha\Psi_\beta^{\beta'}u_{\hat\imath}\underline{\partial}_{\beta'}u_{\hat\jmath}\right), \quad sZ^{I^\dagger}\left(\partial_\alpha\Psi_\beta^{\beta'}v_{\tilde\imath}\underline{\partial}_{\beta'}u_{\hat\jmath}\right),$$
$$sZ^{I^\dagger}\left(\partial_\alpha\Psi_\beta^{\beta'}\partial_\gamma u_{\hat\imath}\underline{\partial}_{\beta'}u_{\hat\jmath}\right), \quad sZ^{I^\dagger}\left(\partial_\alpha\Psi_\beta^{\beta'}\partial_\gamma v_{\tilde\imath}\underline{\partial}_{\beta'}u_{\hat\jmath}\right).$$

We will use the additional decay supplied by $|Z^I\partial_\alpha\Psi_\beta^{\beta'}| \leqslant C(I)t^{-1}$. Let us take

$$Z^{I^\dagger}\left(\partial_\alpha\Psi_\beta^{\beta'}u_{\hat\imath}\underline{\partial}_{\beta'}u_{\hat\jmath}\right)$$

as an example and write its estimate in details:

$$\begin{aligned}
&\left\|sZ^{I^\dagger}\left(\partial_\alpha\Psi_\beta^{\beta'}u_{\hat\imath}\underline{\partial}_{\beta'}u_{\hat\jmath}\right)\right\|_{L^2(\mathcal{H}_s)} \leqslant \sum_{I_1+I_2+I_3=I^\dagger}\left\|sZ^{I_3}\partial_\alpha\Psi_\beta^{\beta'}\,Z^{I_1}u_{\hat\imath}\,Z^{I_2}\underline{\partial}_{\beta'}u_{\hat\jmath}\right\|_{L^2(\mathcal{H}_s)}\\
&\leqslant C\sum_{|I_1|+|I_2|\leqslant|I^\dagger|}\left\|st^{-1}Z^{I_1}u_{\hat\imath}\,Z^{I_2}\underline{\partial}_{\beta'}u_{\hat\jmath}\right\|_{L^2(\mathcal{H}_s)}\\
&\leqslant C\sum_{|I_2|\leqslant|I^\dagger|}\left\|st^{-1}u_{\hat\imath}\,Z^{I_2}\underline{\partial}_{\beta'}u_{\hat\jmath}\right\|_{L^2(\mathcal{H}_s)} + C\sum_{\substack{|I_1|=1\\|I_2|\leqslant|I^\dagger|-1}}\left\|st^{-1}Z^{I_1}u_{\hat\imath}\,Z^{I_2}\underline{\partial}_{\beta'}u_{\hat\jmath}\right\|_{L^2(\mathcal{H}_s)}\\
&\quad + C\sum_{\substack{|I_1|=2\\|I_2|\leqslant|I^\dagger|-2}}\left\|st^{-1}Z^{I_1}u_{\hat\imath}\,Z^{I_2}\underline{\partial}_{\beta'}u_{\hat\jmath}\right\|_{L^2(\mathcal{H}_s)}\\
&\quad + C\sum_{\substack{|I_1|=3\\|I_2|\leqslant|I^\dagger|-3}}\left\|st^{-1}Z^{I_1}u_{\hat\imath}\,Z^{I_2}\underline{\partial}_{\beta'}u_{\hat\jmath}\right\|_{L^2(\mathcal{H}_s)} + C\sum_{|I_1|=4}\left\|st^{-1}Z^{I_1}u_{\hat\imath}\,\underline{\partial}_{\beta'}u_{\hat\jmath}\right\|_{L^2(\mathcal{H}_s)}\\
&=: T_0 + T_1 + T_2 + T_3 + T_4.
\end{aligned}$$

Table 4

Terms	$(4, \leqslant 0)$	$(3, \leqslant 1)$	$(2, \leqslant 2)$	$(1, \leqslant 3)$	$(0, \leqslant 4)$
$u_{\hat{\imath}}\underline{\partial}_a\underline{\partial}_\beta u_{\hat{\jmath}}$	(7.3.6b), (7.5.4c)	(7.5.5a), (7.3.5c)	(7.5.5a), (7.3.5c)	(7.5.5c), (7.3.5a)	(7.5.5c), (7.3.5a)
$v_{\check{\imath}}\underline{\partial}_a\underline{\partial}_\beta u_{\hat{\jmath}}$	(7.3.1e), (7.5.4c)	(7.5.1e), (7.3.5c)	(7.5.1e), (7.3.5c)	(7.5.1e), (7.3.5a)	(7.5.1e), (7.3.5a)
$u_{\hat{\imath}}\partial_t\underline{\partial}_a u_{\hat{\jmath}}$	(7.3.6b), (7.5.4c)	(7.5.5a), (7.3.5c)	(7.5.5a), (7.3.5c)	(7.5.5c), (7.3.5a)	(7.5.5c), (7.3.5a)
$v_{\check{\imath}}\partial_t\underline{\partial}_a u_{\hat{\jmath}}$	(7.3.1e), (7.5.4c)	(7.5.1e), (7.3.5c)	(7.5.1e), (7.3.5c)	(7.5.1e), (7.3.5a)	(7.5.1e), (7.3.5a)
$\underline{\partial}_\gamma u_{\hat{\imath}}\underline{\partial}_a\underline{\partial}_\beta u_{\hat{\jmath}}$	(7.3.1a), (7.5.4c)	(7.5.1a), (7.3.5c)	(7.5.1a), (7.3.5c)	(7.5.3a), (7.3.5a)	(7.5.3a), (7.3.5a)
$\underline{\partial}_\gamma v_{\check{\imath}}\underline{\partial}_a\underline{\partial}_\beta u_{\hat{\jmath}}$	(7.3.2a), (7.5.4c)	(7.5.2a), (7.3.5c)	(7.5.2a), (7.3.5c)	(7.5.2a), (7.3.5a)	(7.5.2a), (7.3.5a)

Table 5

Terms	$(4, \leqslant 0)$	$(3, \leqslant 1)$	$(2, \leqslant 2)$	$(1, \leqslant 3)$	$(0, \leqslant 4)$
$\underline{\partial}_\gamma u_{\hat{\imath}}\partial_t\underline{\partial}_a u_{\hat{\jmath}}$	(7.3.1a), (7.5.4c)	(7.5.1a), (7.3.5c)	(7.5.1a), (7.3.5c)	(7.5.3a), (7.3.5a)	(7.5.3a), (7.3.5a)
$\underline{\partial}_\gamma v_{\check{\imath}}\partial_t\underline{\partial}_a u_{\hat{\jmath}}$	(7.3.2a), (7.5.4c)	(7.5.2a), (7.3.5c)	(7.5.2a), (7.3.5c)	(7.5.2a), (7.3.5a)	(7.5.2a), (7.3.5a)
$v_{\check{\imath}}\partial_\alpha\partial_\beta v_{\check{\jmath}}$	(7.3.1e), (7.5.2a)	(7.5.1e), (7.3.2a)	(7.5.1e), (7.3.2a)	(7.5.1e), (7.3.2a)	(7.5.1e), (7.3.2a)
$\partial_\gamma u_{\hat{\imath}}\partial_\alpha\partial_\beta v_{\check{\jmath}}$	(7.3.1a), (7.5.2a)	(7.5.1a), (7.3.2a)	(7.5.1a), (7.3.2a)	(7.5.3a), (7.3.2a)	(7.5.3a), (7.3.2a)
$\partial_\gamma v_{\check{\imath}}\partial_\alpha\partial_\beta v_{\check{\jmath}}$	(7.3.2a), (7.5.2a)	(7.5.2a), (7.3.2a)	(7.5.2a), (7.3.2a)	(7.5.2a), (7.3.2a)	(7.5.2a), (7.3.2a)

For the term T_0 by (7.5.5c) and (7.3.1a), we have

$$\begin{aligned}
T_0 =& C \sum_{|I_2| \leqslant |I^\dagger|} \big\| s t^{-1} u_{\hat\imath}\, Z^{I_2} \underline{\partial}_{\beta'} u_{\hat\jmath} \big\|_{L^2(\mathcal{H}_s)} \\
\leqslant& C C_1 \epsilon \sum_{|I_2| \leqslant |I^\dagger|} \big\| s t^{-1} t^{-3/2} s\, Z^{I_2} \underline{\partial}_{\beta'} u_{\hat\jmath} \big\|_{L^2(\mathcal{H}_s)} \\
=& C C_1 \epsilon \sum_{|I_2| \leqslant |I^\dagger|} \big\| t^{-5/2} s^2\, Z^{I_2} \underline{\partial}_{\beta'} u_{\hat\jmath} \big\|_{L^2(\mathcal{H}_s)} \\
\leqslant& C C_1 \epsilon s^{-1/2} \sum_{|I_2| \leqslant |I^\dagger|} \big\| Z^{I_2} \underline{\partial}_{\beta'} u_{\hat\jmath} \big\|_{L^2(\mathcal{H}_s)} \\
\leqslant& C (C_1 \epsilon)^2 s^{-1/2+\delta}.
\end{aligned}$$

The term T_1 is estimated by (7.5.5c) and (7.3.3a):

$$\begin{aligned}
T_1 =& C \sum_{\substack{|I_1|=1 \\ |I_2| \leqslant |I^\dagger|-1}} \big\| s t^{-1} Z^{I_1} u_{\hat\imath}\, Z^{I_2} \underline{\partial}_{\beta'} u_{\hat\jmath} \big\|_{L^2(\mathcal{H}_s)} \\
\leqslant& C C_1 \epsilon \sum_{\substack{|I_1|=1 \\ |I_2| \leqslant |I^\dagger|-1}} \big\| s t^{-1} t^{-3/2} s\, Z^{I_2} \underline{\partial}_{\beta'} u_{\hat\jmath} \big\|_{L^2(\mathcal{H}_s)} \\
\leqslant& C C_1 \epsilon s^{-1/2} \sum_{\substack{|I_1|=1 \\ |I_2| \leqslant |I^\dagger|-1}} \big\| Z^{I_2} \underline{\partial}_{\beta'} u_{\hat\jmath} \big\|_{L^2(\mathcal{H}_s)} \\
\leqslant& C (C_1 \epsilon)^2 s^{-1/2+\delta}.
\end{aligned}$$

The term T_2 is estimated by (7.5.5a) and (7.3.3a):

$$\begin{aligned}
T_2 =& C \sum_{\substack{|I_1|=2 \\ |I_2| \leqslant |I^\dagger|-2}} \big\| s t^{-1} Z^{I_1} u_{\hat\imath}\, Z^{I_2} \underline{\partial}_{\beta'} u_{\hat\jmath} \big\|_{L^2(\mathcal{H}_s)} \\
\leqslant& C C_1 \epsilon \sum_{\substack{|I_1|=2 \\ |I_2| \leqslant |I^\dagger|-2}} \big\| s t^{-1} t^{-3/2} s^{1+\delta}\, Z^{I_2} \underline{\partial}_{\beta'} u_{\hat\jmath} \big\|_{L^2(\mathcal{H}_s)} \\
\leqslant& C C_1 \epsilon s^{-1/2+\delta} \sum_{\substack{|I_1|=2 \\ |I_2| \leqslant |I^\dagger|-2}} \big\| Z^{I_2} \underline{\partial}_{\beta'} u_{\hat\jmath} \big\|_{L^2(\mathcal{H}_s)} \\
\leqslant& C (C_1 \epsilon)^2 s^{-1/2+\delta}.
\end{aligned}$$

The term T_3 is estimated by (7.5.5a) and (7.3.3a):

$$
\begin{aligned}
T_3 =& C \sum_{\substack{|I_1|=3\\|I_1|\leqslant|I^\dagger|-3}} \left\|st^{-1}Z^{I_1}u_{\hat{\imath}}\, Z^{I_2}\underline{\partial}_{\beta'}u_{\hat{\jmath}}\right\|_{L^2(\mathcal{H}_s)}\\
\leqslant& CC_1\epsilon \sum_{\substack{|I_1|=3\\|I_2|\leqslant|I^\dagger|-3}} \left\|st^{-1}t^{-3/2}s^{1+\delta}\, Z^{I_2}\underline{\partial}_{\beta'}u_{\hat{\jmath}}\right\|_{L^2(\mathcal{H}_s)}\\
\leqslant& CC_1\epsilon s^{-1/2+\delta} \sum_{\substack{|I_1|=3\\|I_2|\leqslant|I^\dagger|-3}} \left\|Z^{I_2}\underline{\partial}_{\beta'}u_{\hat{\jmath}}\right\|_{L^2(\mathcal{H}_s)}\\
\leqslant& C(C_1\epsilon)^2 s^{-1/2+\delta}
\end{aligned}
$$

The term T_4 is estimated by (7.3.6a) and (7.5.3a):

$$
\begin{aligned}
T_4 =& C \sum_{|I_2|=4} \left\|st^{-1}Z^{I_2}u_{\hat{\imath}}\, \underline{\partial}_{\beta'}u_{\hat{\jmath}}\right\|_{L^2(\mathcal{H}_s)}\\
\leqslant& CC_1\epsilon \sum_{|I_2|=4} \left\|st^{-1}Z^{I_2}u_{\hat{\imath}}\, t^{-1/2}s^{-1}\right\|_{L^2(\mathcal{H}_s)}\\
=& CC_1\epsilon s^{-1/2} \sum_{|I_2|=4} \left\|t^{-1}Z^{I_2}u_{\hat{\imath}}\right\|_{L^2(\mathcal{H}_s)}\\
\leqslant& C(C_1\epsilon)^2 s^{-1/2+\delta}.
\end{aligned}
$$

For the remaining three terms, we list out the L^2 and L^∞ estimates to be used for each term and each partition of $I^\dagger$, recall here the notation $(a, \leqslant b)$ means $|I_1| = a, |I_2| \leqslant b$: see Table 6. Finally we conclude with (8.5.1a).

We estimate the term $Q_{T\hat{\imath}}$ which is similar to that of $Q_{G\hat{\imath}}$. Recall the structure of $Q_{T\hat{\imath}}$ presented in (8.3.1). We take the term $Z^{I_2}\partial_\gamma u_{\hat{k}}\, \underline{\partial}_a\underline{\partial}_\beta Z^{I_1}u_{\hat{\jmath}}$ as an example:

$$
\begin{aligned}
&\sum_{|I_1|+|I_2|\leqslant|I^\dagger|} \left\|sZ^{I_2}\partial_\gamma u_{\hat{k}}\, \underline{\partial}_a\underline{\partial}_\beta Z^{I_1}u_{\hat{\jmath}}\right\|_{L^2(\mathcal{H}_s)}\\
=& \sum_{|I_2|\leqslant|I^\dagger|} \left\|sZ^{I_2}\partial_\gamma u_{\hat{k}}\, \underline{\partial}_a\underline{\partial}_\beta u_{\hat{\jmath}}\right\|_{L^2(\mathcal{H}_s)} + \sum_{\substack{|I_1|+|I_2|\leqslant|I^\dagger|\\|I_1|=1}} \left\|sZ^{I_2}\partial_\gamma u_{\hat{k}}\, \underline{\partial}_a\underline{\partial}_\beta Z^{I_1}u_{\hat{\jmath}}\right\|_{L^2(\mathcal{H}_s)}\\
&+ \sum_{\substack{|I_1|+|I_2|\leqslant|I^\dagger|\\|I_1|=2}} \left\|sZ^{I_2}\partial_\gamma u_{\hat{k}}\, \underline{\partial}_a\underline{\partial}_\beta Z^{I_1}u_{\hat{\jmath}}\right\|_{L^2(\mathcal{H}_s)}\\
&+ \sum_{\substack{|I_1|+|I_2|\leqslant|I^\dagger|\\|I_1|=3}} \left\|sZ^{I_2}\partial_\gamma u_{\hat{k}}\, \underline{\partial}_a\underline{\partial}_\beta Z^{I_1}u_{\hat{\jmath}}\right\|_{L^2(\mathcal{H}_s)}\\
&+ \sum_{|I_1|\leqslant|I^\dagger|} \left\|s\partial_\gamma u_{\hat{k}}\, \underline{\partial}_a\underline{\partial}_\beta Z^{I_1}u_{\hat{\jmath}}\right\|_{L^2(\mathcal{H}_s)}\\
=:& T_0 + T_1 + T_2 + T_3 + T_4.
\end{aligned}
$$

Table 6

Terms	$(4, \leqslant 0)$	$(3, \leqslant 1)$	$(2, \leqslant 2)$	$(1, \leqslant 3)$	$(0, \leqslant 4)$
$u_{\hat{\imath}}\underline{\partial}_{\beta'}u_{\hat{\jmath}}$	(7.3.6b), (7.5.3a)	(7.5.5a), (7.3.3a)	(7.5.5a), (7.3.3a)	(7.5.5c), (7.3.3a)	(7.5.5c), (7.3.1a)
$v_{\check{\imath}}\underline{\partial}_{\beta'}u_{\hat{\jmath}}$	(7.3.1e), (7.5.3a)	(7.5.1e), (7.3.3a)	(7.5.1e), (7.3.3a)	(7.5.1e), (7.3.3a)	(7.5.1e), (7.3.1a)
$\partial_\gamma u_{\hat{\imath}}\underline{\partial}_{\beta'}u_{\hat{\jmath}}$	(7.3.1a), (7.5.3a)	(7.5.1a), (7.3.3a)	(7.5.1a), (7.3.3a)	(7.5.3a), (7.3.3a)	(7.5.3a), (7.3.1a)
$\partial_\gamma v_{\check{\imath}}\underline{\partial}_{\beta'}u_{\hat{\jmath}}$	(7.3.2a), (7.5.3a)	(7.5.2a), (7.3.3a)	(7.5.2a), (7.3.3a)	(7.5.2a), (7.3.3a)	(7.5.2a), (7.3.1a)

Table 7

Products	$(4, \leqslant 0)$	$(3, \leqslant 1)$	$(2, \leqslant 2)$	$(1, \leqslant 3)$	$(0, \leqslant 4)$
$\partial_\alpha u_{\hat{\jmath}}\partial_\beta u_{\hat{k}}$	(7.3.1a), (7.5.3a)	(7.3.3a), (7.5.3a)	(7.5.1a), (7.3.3a)	(7.5.3a), (7.3.3a)	(7.5.3a), (7.3.1a)
$\partial_\alpha v_{\check{\jmath}}\partial_\beta w_k$	(7.3.2a), (7.5.1a)	(7.3.2a), (7.5.1a)	(7.5.2a), (7.3.1a)	(7.5.2a), (7.3.1a)	(7.5.2a), (7.3.1a)
$v_{\hat{k}}\partial_\alpha w_j$	(7.3.1e), (7.5.1a)	(7.3.1e), (7.5.1a)	(7.5.1e), (7.3.1a)	(7.5.1e), (7.3.1a)	(7.5.1e), (7.3.1a)
$v_{\hat{k}}v_{\hat{\jmath}}$	(7.3.1e), (7.5.1e)	(7.3.1e), (7.5.1e)	(7.5.1e), (7.3.1e)	(7.5.1e), (7.3.1e)	(7.5.1e), (7.3.1e)

We will estimate each partition of $I^\dagger$:

$$
\begin{aligned}
T_0 &= \sum_{|I_2|\leqslant|I^\dagger|} \left\|sZ^{I_2}\partial_\gamma u_{\hat{k}}\,\underline{\partial}_a\underline{\partial}_\beta u_{\hat{\jmath}}\right\|_{L^2(\mathcal{H}_s)} \leqslant \sum_{|I_2|\leqslant 4} \left\|sZ^{I_2}\partial_\gamma u_{\hat{k}}\,\underline{\partial}_a\underline{\partial}_\beta u_{\hat{\jmath}}\right\|_{L^2(\mathcal{H}_s)}\\
&\leqslant \sum_{|I_2|\leqslant 4} \left\|(s/t)Z^{I_2}\partial_\gamma u_{\hat{k}}\right\|_{L^2(\mathcal{H}_s)} \left\|s(t/s)\underline{\partial}_a\underline{\partial}_\beta u_{\hat{\jmath}}\right\|_{L^\infty(\mathcal{H}_s)} \leqslant CC_1\epsilon s^\delta\, CC_1\epsilon s^{-3/2}\\
&\leqslant C(C_1\epsilon)^2 s^{-3/2+\delta},
\end{aligned}
$$

where (7.3.1a) and (7.4.5c) are used.

$$
\begin{aligned}
T_1 &\leqslant \sum_{|I_1|=1,|I_2|\leqslant 3} \left\|sZ^{I_2}\partial_\gamma u_{\hat{k}}\,\underline{\partial}_a\underline{\partial}_\beta Z^{I_1} u_{\hat{\jmath}}\right\|_{L^2(\mathcal{H}_s)}\\
&\leqslant \sum_{|I_1|=1,|I_2|\leqslant 3} \left\|(s/t)Z^{I_2}\partial_\gamma u_{\hat{k}}\right\|_{L^2(\mathcal{H}_s)}\left\|s(t/s)\underline{\partial}_a\underline{\partial}_\beta Z^{I_1} u_{\hat{\jmath}}\right\|_{L^\infty(\mathcal{H}_s)}\\
&\leqslant CC_1\epsilon\, CC_1\epsilon s^{-3/2+\delta} \leqslant C(C_1\epsilon)^2 s^{-3/2+\delta},
\end{aligned}
$$

where (7.3.3a) and (7.4.5a) are used.

Similarly, the term T_2, T_3 and T_4 are estimated by applying respectively (7.3.3a)(7.4.5a), (7.3.4a), (7.5.3a), (7.3.4a), and (7.5.3a).

For the remaining terms, we will not write in details the proof but list out the inequalities to be used on each term and each partition of the index $|I_1| + |I_2| \leqslant |I^\dagger|$, in the following list:

Products	$(4, \leqslant 0)$	$(3, \leqslant 1)$	$(2, \leqslant 2)$
$s\underline{\partial}_a\underline{\partial}_\beta Z^{I_1}u_{\hat{\imath}}Z^{I_2}\partial_\gamma u_{\hat{k}}$	(7.3.4a), (7.5.3a)	(7.3.4a), (7.5.3a)	(7.3.4c), (7.5.1a)
$s\underline{\partial}_a\underline{\partial}_\beta Z^{I_1}u_{\hat{\imath}}Z^{I_2}\partial_\gamma v_{\check{k}}$	(7.3.4a), (7.5.2a)	(7.3.4a), (7.5.2a)	(7.3.4c), (7.5.2a)
$s\underline{\partial}_a\underline{\partial}_\beta Z^{I_1}u_{\hat{\imath}}Z^{I_2}u_{\hat{k}}$	(7.3.4a), (7.5.5c)	(7.3.4a), (7.5.5c)	(7.3.4c), (7.5.5a)
$s\underline{\partial}_a\underline{\partial}_\beta Z^{I_1}u_{\hat{\imath}}Z^{I_2}v_{\check{k}}$	(7.3.4a), (7.5.1e)	(7.3.4a), (7.5.1e)	(7.3.4c), (7.5.1e)

Products	$(1, \leqslant 3)$	$(0, \leqslant 4)$
$s\underline{\partial}_a\underline{\partial}_\beta Z^{I_1}u_{\hat{\imath}}Z^{I_2}\partial_\gamma u_{\hat{k}}$	(7.4.5a), (7.3.3a)	(7.4.5c), (7.3.1a)
$s\underline{\partial}_a\underline{\partial}_\beta Z^{I_1}u_{\hat{\imath}}Z^{I_2}\partial_\gamma v_{\check{k}}$	(7.4.5a), (7.3.2a)	(7.4.5c), (7.3.2a)
$s\underline{\partial}_a\underline{\partial}_\beta Z^{I_1}u_{\hat{\imath}}Z^{I_2}u_{\hat{k}}$	(7.4.5a), (7.3.6c)	(7.4.5c), (7.3.6a)
$s\underline{\partial}_a\underline{\partial}_\beta Z^{I_1}u_{\hat{\imath}}Z^{I_2}v_{\check{k}}$	(7.4.5a), (7.3.1e)	(7.4.5c), (7.3.1e)

Products	$(4, \leqslant 0)$	$(3, \leqslant 1)$	$(2, \leqslant 2)$
$st^{-1}\partial_{\gamma'}Z^{I_1}u_{\hat{\jmath}}Z^{I_2}\partial_\gamma u_{\hat{k}}$	(7.2.1a), (7.5.3a)	(7.2.3a), (7.5.3a)	(7.2.3a), (7.5.1a)
$st^{-1}\partial_{\gamma'}Z^{I_1}u_{\hat{\jmath}}Z^{I_2}\partial_\gamma v_{\check{k}}$	(7.2.1a), (7.5.2a)	(7.2.3a), (7.5.2a)	(7.2.3a), (7.5.2a)
$st^{-1}\partial_{\gamma'}Z^{I_1}u_{\hat{\jmath}}Z^{I_2}u_{\hat{k}}$	(7.2.1a), (7.5.5c)	(7.2.3a), (7.5.5c)	(7.2.3a), (7.5.5a)
$st^{-1}\partial_{\gamma'}Z^{I_1}u_{\hat{\jmath}}Z^{I_2}v_{\check{k}}$	(7.2.1a), (7.5.1e)	(7.2.3a), (7.5.1e)	(7.2.3a), (7.5.1e)

Products	$(1, \leqslant 3)$	$(0, \leqslant 4)$
$st^{-1}\partial_{\gamma'}Z^{I_1}u_{\hat{\jmath}}Z^{I_2}\partial_\gamma u_{\hat{k}}$	(7.4.4a), (7.3.3a)	(7.4.4a), (7.3.1a)
$st^{-1}\partial_{\gamma'}Z^{I_1}u_{\hat{\jmath}}Z^{I_2}\partial_\gamma v_{\check{k}}$	(7.4.4a), (7.3.2a)	(7.4.4a), (7.3.2a)
$st^{-1}\partial_{\gamma'}Z^{I_1}u_{\hat{\jmath}}Z^{I_2}u_{\hat{k}}$	(7.4.4a), (7.3.6c)	(7.4.4a), (7.3.6a)
$st^{-1}\partial_{\gamma'}Z^{I_1}u_{\hat{\jmath}}Z^{I_2}v_{\check{k}}$	(7.4.4a), (7.3.1e)	(7.4.4a), (7.3.1e)

Now, we estimate the term $F_{\hat{\imath}}$ which is also similar: recall the structure of $Z^{I^\dagger}F_{\hat{\imath}}$ presented in (8.3.3) with I replaced by $I^\dagger$. As before, we consider the term $Z^{I^\dagger}\big(\partial_\alpha u_{\hat{\jmath}}\partial_\beta u_{\hat{k}}\big)$ as an example and we write down the details of the analysis. For the rest terms, we just give the L^2 and L^∞ estimates to be used for each factor:

$$
\begin{aligned}
&\big\|sZ^{I^\dagger}\big(\partial_\alpha u_{\hat{\jmath}}\partial_\beta u_{\hat{k}}\big)\big\|_{L^2(\mathcal{H}_s)}\\
&\leqslant \big\|s\partial_\alpha u_{\hat{\jmath}}\,Z^{I^\dagger}\partial_\beta u_{\hat{k}}\big\|_{L^2(\mathcal{H}_s)} + \sum_{\substack{I_1+I_2=I^\dagger\\|I_1|=1}}\big\|sZ^{I_1}\partial_\alpha u_{\hat{\jmath}}\,Z^{I_2}\partial_\beta u_{\hat{k}}\big\|_{L^2(\mathcal{H}_s)}\\
&\quad+\sum_{\substack{I_1+I_2=I^\dagger\\|I_1|=2}}\big\|sZ^{I_1}\partial_\alpha u_{\hat{\jmath}}\,Z^{I_2}\partial_\beta u_{\hat{k}}\big\|_{L^2(\mathcal{H}_s)}\\
&\quad+\sum_{\substack{I_1+I_2=I^\dagger\\|I_1|=3}}\big\|sZ^{I_1}\partial_\alpha u_{\hat{\jmath}}\,Z^{I_2}\partial_\beta u_{\hat{k}}\big\|_{L^2(\mathcal{H}_s)} + \big\|sZ^{I^\dagger}\partial_\alpha u_{\hat{\jmath}}\,\partial_\beta u_{\hat{k}}\big\|_{L^2(\mathcal{H}_s)}\\
&=: T_0+T_1+T_2+T_3+T_4.
\end{aligned}
$$

The term T_0 is estimated by (7.5.3a) and (7.3.1a):

$$
\begin{aligned}
T_0 =&\big\|s\partial_\alpha u_{\hat{\jmath}}\,Z^{I^\dagger}\partial_\beta u_{\hat{k}}\big\|_{L^2(\mathcal{H}_s)} \leqslant CC_1\epsilon\big\|s\,t^{-1/2}s^{-1}\,Z^{I^\dagger}\partial_\beta u_{\hat{k}}\big\|_{L^2(\mathcal{H}_s)}\\
=&CC_1\epsilon\big\|s\,t^{-1/2}s^{-1}\,(t/s)\big(s/tZ^{I^\dagger}\partial_\beta u_{\hat{k}}\big)\big\|_{L^2(\mathcal{H}_s)} \leqslant CC_1\epsilon\big\|s/tZ^{I^\dagger}\partial_\beta u_{\hat{k}}\big\|_{L^2(\mathcal{H}_s)}\\
\leqslant& C(C_1\epsilon)^2 s^\delta.
\end{aligned}
$$

Here the relation $t \leqslant Cs^2$ (in $\mathcal{K}$) is used. The term T_4 is estimated in the same way by exchanging the role of $u_{\hat{\jmath}}$ and $u_{\hat{k}}$.

The terms T_1 and T_3 are estimated by (7.5.3a) and (7.3.3a). We estimate T_1 as follows. The estimate of T_3 is done by exchanging $u_{\hat{\jmath}}$ and $u_{\hat{k}}$ in the following argument:

$$
\begin{aligned}
T_1 =& \sum_{\substack{I_1+I_2=I^\dagger\\|I_1|=1}}\big\|sZ^{I_1}\partial_\alpha u_{\hat{\jmath}}\,Z^{I_2}\partial_\beta u_{\hat{k}}\big\|_{L^2(\mathcal{H}_s)}\\
\leqslant& CC_1\epsilon\big\|s\,t^{-1/2}s^{-1}\,(t/s)(s/t)Z^{I_2}\partial_\beta u_{\hat{k}}\big\|_{L^2(\mathcal{H}_s)}\\
\leqslant& CC_1\epsilon\big\|(s/t)Z^{I_2}\partial_\beta u_{\hat{k}}\big\|_{L^2(\mathcal{H}_s)} \leqslant C(C_1\epsilon)^2.
\end{aligned}
$$

Table 8

Products	$(3, \leqslant 0)$	$(2, \leqslant 1)$	$(1, \leqslant 2)$	$(0, \leqslant 3)$
$u_{\hat\imath}\underline{\partial}_a\underline{\partial}_\beta u_{\hat\jmath}$	(7.3.6c), (7.5.4c)	(7.3.6c), (7.5.4b)	(7.5.5c), (7.3.5c)	(7.5.5c), (7.3.5b)
$v_{\check\imath}\underline{\partial}_a\underline{\partial}_\beta u_{\hat\jmath}$	(7.3.1f), (7.5.4c)	(7.3.1f), (7.5.4b)	(7.5.1f), (7.3.5c)	(7.5.1f), (7.3.5b)
$u_{\hat\imath}\partial_t\underline{\partial}_a u_{\hat\jmath}$	(7.3.6c), (7.5.4c)	(7.3.6c), (7.5.4b)	(7.5.5c), (7.3.5c)	(7.5.5c), (7.3.5b)
$v_{\check\imath}\partial_t\underline{\partial}_a u_{\hat\jmath}$	(7.3.1f), (7.5.4c)	(7.3.1f), (7.5.4b)	(7.5.1f), (7.3.5c)	(7.5.1f), (7.3.5b)
$\underline{\partial}_\gamma u_{\hat\imath}\underline{\partial}_a\underline{\partial}_\beta u_{\hat\jmath}$	(7.3.3a), (7.5.4c)	(7.3.3a), (7.5.4b)	(7.5.3a), (7.3.5c)	(7.5.3a), (7.3.5b)
$\underline{\partial}_\gamma v_{\check\imath}\underline{\partial}_a\underline{\partial}_\beta u_{\hat\jmath}$	(7.3.2b), (7.5.4c)	(7.3.2b), (7.5.4b)	(7.5.2b), (7.3.5c)	(7.5.2b), (7.3.5b)

Table 9

Products	$(3, \leqslant 0)$	$(2, \leqslant 1)$	$(1, \leqslant 2)$	$(0, \leqslant 3)$
$\underline{\partial}_\gamma u_{\hat\imath}\partial_t\underline{\partial}_a u_{\hat\jmath}$	(7.3.3a), (7.5.4c)	(7.3.3a), (7.5.4b)	(7.5.3a), (7.3.5c)	(7.5.3a), (7.3.5b)
$\underline{\partial}_\gamma v_{\check\imath}\partial_t\underline{\partial}_a u_{\hat\jmath}$	(7.3.2b), (7.5.4c)	(7.3.2b), (7.5.4b)	(7.5.2b), (7.3.5c)	(7.5.2b), (7.3.5b)
$v_{\check\imath}\partial_\alpha\partial_\beta v_{\check\jmath}$	(7.3.1f), (7.5.2b)	(7.3.1f), (7.5.2b)	(7.5.1f), (7.3.2b)	(7.5.1f), (7.3.2a)
$\partial_\gamma u_{\hat\imath}\partial_\alpha\partial_\beta v_{\check\jmath}$	(7.3.3a), (7.5.2b)	(7.3.3a), (7.5.2b)	(7.5.3a), (7.3.2b)	(7.5.3a), (7.3.2a)
$\partial_\gamma v_{\check\imath}\partial_\alpha\partial_\beta v_{\check\jmath}$	(7.3.2b), (7.5.2b)	(7.3.2b), (7.5.2b)	(7.5.2b), (7.3.2b)	(7.5.2b), (7.3.2a)

Table 10

Terms	$(3, \leqslant 0)$	$(2, \leqslant 1)$	$(1, \leqslant 2)$	$(0, \leqslant 3)$
$\partial_\alpha \Psi_\beta^{\beta'} u_{\hat\imath} \underline{\partial}_{\beta'} u_{\hat\jmath}$	(7.3.6c), (7.5.3a)	(7.3.6c), (7.5.3a)	(7.5.5c), (7.3.3a)	(7.5.5c), (7.3.3a)
$\partial_\alpha \Psi_\beta^{\beta'} v_{\check\imath} \underline{\partial}_{\beta'} u_{\hat\jmath}$	(7.3.1f), (7.5.3a)	(7.3.1f), (7.5.3a)	(7.5.1f), (7.3.3a)	(7.5.1f), (7.3.3a)
$\partial_\alpha \Psi_\beta^{\beta'} \partial_\gamma u_{\hat\imath} \underline{\partial}_{\beta'} u_{\hat\jmath}$	(7.3.3a), (7.5.3a)	(7.3.3a), (7.5.3a)	(7.5.3a), (7.3.3a)	(7.5.3a), (7.3.3a)
$\partial_\alpha \Psi_\beta^{\beta'} \partial_\gamma v_{\check\imath} \underline{\partial}_{\beta'} u_{\hat\jmath}$	(7.3.2b), (7.5.3a)	(7.3.2b), (7.5.3a)	(7.5.2b), (7.3.3a)	(7.5.2b), (7.3.3a)

Table 11

Products	$(3, \leqslant 0)$	$(2, \leqslant 1)$	$(1, \leqslant 2)$	$(0, \leqslant 3)$
$s\underline{\partial}_a \underline{\partial}_\beta Z^{J_1} u_{\hat\imath} Z^{J_2} \partial_\gamma u_{\hat k}$	(7.3.4b), (7.5.3a)	(7.3.4c), (7.5.3a)	(7.4.5b), (7.3.3a)	(7.4.5c), (7.3.3a)
$s\underline{\partial}_a \underline{\partial}_\beta Z^{J_1} u_{\hat\imath} Z^{J_2} \partial_\gamma v_{\check k}$	(7.3.4b), (7.5.2b)	(7.3.4c), (7.5.2b)	(7.4.5b), (7.3.2b)	(7.4.5c), (7.3.2b)
$s\underline{\partial}_a \underline{\partial}_\beta Z^{J_1} u_{\hat\imath} Z^{J_2} u_{\hat k}$	(7.3.4b), (7.5.5c)	(7.3.4c), (7.5.5c)	(7.4.5b), (7.3.6c)	(7.4.5c), (7.3.6c)
$s\underline{\partial}_a \underline{\partial}_\beta Z^{J_1} u_{\hat\imath} Z^{J_2} v_{\check k}$	(7.3.4b), (7.5.1f)	(7.3.4c), (7.5.1f)	(7.4.5b), (7.3.1f)	(7.4.5c), (7.3.1f)

Table 12

Products	$(3, \leqslant 0)$	$(2, \leqslant 1)$	$(1, \leqslant 2)$	$(0, \leqslant 3)$
$st^{-1}\partial_{\gamma'} Z^{J_1} u_{\hat\jmath} Z^{J_2} \partial_\gamma u_{\hat k}$	(7.2.3a), (7.5.3a)	(7.2.3a), (7.5.3a)	(7.4.4a), (7.3.3a)	(7.4.4a), (7.3.3a)
$st^{-1}\partial_{\gamma'} Z^{J_1} u_{\hat\jmath} Z^{J_2} \partial_\gamma v_{\check k}$	(7.2.3a), (7.5.2b)	(7.2.3a), (7.5.2b)	(7.4.4a), (7.3.2b)	(7.4.4a), (7.3.2b)
$st^{-1}\partial_{\gamma'} Z^{J_1} u_{\hat\jmath} Z^{J_2} u_{\hat k}$	(7.2.3a), (7.5.5c)	(7.2.3a), (7.5.5c)	(7.4.4a), (7.3.6c)	(7.4.4a), (7.3.6c)
$st^{-1}\partial_{\gamma'} Z^{J_1} u_{\hat\jmath} Z^{J_2} v_{\check k}$	(7.2.3a), (7.5.1f)	(7.2.3a), (7.5.1f)	(7.4.4a), (7.3.1f)	(7.4.4a), (7.3.1f)

The term T_2 is estimated by (7.5.1a) and (7.3.3a):

$$
\begin{aligned}
T_2 = & \sum_{\substack{I_1+I_2=I^\dagger \\ |I_1|=2}} \left\| s Z^{I_1}\partial_\alpha u_{\hat{\jmath}} Z^{I_2}\partial_\beta u_{\hat{k}} \right\|_{L^2(\mathcal{H}_s)} \\
\leqslant & C C_1\epsilon \left\| s\, t^{-1/2} s^{-1+\delta}\,(t/s)(s/t) Z^{I_2}\partial_\beta u_{\hat{k}} \right\|_{L^2(\mathcal{H}_s)} \\
= & C C_1 \epsilon s^{\delta} \left\| (s/t) Z^{I_2}\partial_\beta u_{\hat{k}} \right\|_{L^2(\mathcal{H}_s)} \leqslant C(C_1\epsilon)^2 s^{\delta}.
\end{aligned}
$$

The estimate of other terms are presented in Table 7. We conclude (8.5.1c).

The estimate of $R(Z^I u_{\hat{\imath}})$ is a direct result of (7.3.4a). □

Proof of Lemma 8.5.2. The proof is essentially the same as the one of Lemma 8.5.1. The main difference lies in the inequalities we use for each term and partition of the index. We will list out the relevant inequalities and skip the details.

For the proof of (8.5.2a), we list out the inequalities in Table 8.

The following four terms are estimated by apply the additional decay rate supplied by $\partial_\alpha \Phi_\beta^\gamma$:

$$
\begin{aligned}
& sZ^I\left(\partial_\alpha \Psi_\beta^{\beta'} u_{\hat{\imath}} \underline{\partial}_{\beta'} u_{\hat{\jmath}}\right), \quad sZ^I\left(\partial_\alpha \Psi_\beta^{\beta'} v_{\check{\imath}} \underline{\partial}_{\beta'} u_{\hat{\jmath}}\right), \\
& sZ^I\left(\partial_\alpha \Psi_\beta^{\beta'} \partial_\gamma u_{\hat{\imath}} \underline{\partial}_{\beta'} u_{\hat{\jmath}}\right), \quad sZ^I\left(\partial_\alpha \Psi_\beta^{\beta'} \partial_\gamma v_{\check{\imath}} \underline{\partial}_{\beta'} u_{\hat{\jmath}}\right).
\end{aligned}
$$

See Table 10.

For the term $Q_{T\hat{\imath}}$, we find Table 11 and Table 12.

For the estimates on $Z^I F_{\hat{\imath}}$, the inequalities we use are presented in the following list:

Products	$(3, \leqslant 0)$	$(2, \leqslant 1)$	$(1, \leqslant 2)$	$(0, \leqslant 3)$
$\partial_\alpha u_{\hat{\jmath}} \partial_\beta u_{\hat{k}}$	(7.3.3a), (7.5.3a)	(7.3.3a), (7.5.3a)	(7.5.3a), (7.3.3a)	(7.5.3a), (7.3.3a)
$\partial_\alpha v_{\check{\jmath}} \partial_\beta w_k$	(7.3.2b), (7.5.1b)	(7.3.2b), (7.5.1b)	(7.5.2b), (7.3.1b)	(7.5.2b), (7.3.1b)
$v_{\check{k}} \partial_\alpha w_j$	(7.3.1f), (7.5.1b)	(7.3.1f), (7.5.1b)	(7.5.1f), (7.3.1b)	(7.5.1f), (7.3.1b)
$v_{\check{k}} v_{\check{\jmath}}$	(7.3.1f), (7.5.1f)	(7.3.1f), (7.5.1f)	(7.5.1f), (7.3.1f)	(7.5.1f), (7.3.1f)

The estimate on $R(Z^I u_{\hat{\imath}})$ is a direct result of (7.3.4b) □

Proof of Lemma 8.5.3. The proof is essentially the same as the one of Lemma 8.5.1. The main difference is the level of regularity under consideration. We will not give the details and only list the inequalities we use for each term and partition $I = I_2 + I_3$:

Products	$(2, \leqslant 0)$	$(1, \leqslant 1)$	$(0, \leqslant 2)$
$u_{\hat{\imath}} \underline{\partial}_a \partial_\beta u_{\hat{\jmath}}$	(7.3.6c), (7.5.4c)	(7.5.5c), (7.3.5c)	(7.5.5c), (7.3.5c)
$v_{\check{\imath}} \underline{\partial}_a \partial_\beta u_{\hat{\jmath}}$	(7.3.1e), (7.5.4c)	(7.5.1e), (7.3.5c)	(7.5.1e), (7.3.5c)
$u_{\hat{\imath}} \partial_t \underline{\partial}_a u_{\hat{\jmath}}$	(7.3.6c), (7.5.4c)	(7.5.5c), (7.3.5c)	(7.5.5c), (7.3.5c)
$v_{\check{\imath}} \partial_t \underline{\partial}_a u_{\hat{\jmath}}$	(7.3.1e), (7.5.4c)	(7.5.1e), (7.3.5c)	(7.5.1e), (7.3.5c)

Products	$(2, \leqslant 0)$	$(1, \leqslant 1)$	$(0, \leqslant 2)$
$\underline{\partial}_\gamma u_{\hat{\imath}} \underline{\partial}_a \partial_\beta u_{\hat{\jmath}}$	(7.3.3a), (7.5.4c)	(7.5.3a), (7.3.5c)	(7.5.3a), (7.3.5c)
$\underline{\partial}_\gamma v_{\check{\imath}} \underline{\partial}_a \partial_\beta u_{\hat{\jmath}}$	(7.3.2a), (7.5.4c)	(7.5.2a), (7.3.5c)	(7.5.2a), (7.3.5c)
$\underline{\partial}_\gamma u_{\hat{\imath}} \partial_t \underline{\partial}_a u_{\hat{\jmath}}$	(7.3.3a), (7.5.4c)	(7.5.3a), (7.3.5c)	(7.5.3a), (7.3.5c)
$\underline{\partial}_\gamma v_{\check{\imath}} \partial_t \underline{\partial}_a u_{\hat{\jmath}}$	(7.3.2a), (7.5.4c)	(7.5.2a), (7.3.5c)	(7.5.2a), (7.3.5c)
$v_{\check{\imath}} \partial_\alpha \partial_\beta v_{\check{\jmath}}$	(7.3.1e), (7.5.2a)	(7.5.1e), (7.3.2a)	(7.5.1e), (7.3.2a)
$\partial_\gamma u_{\hat{\imath}} \partial_\alpha \partial_\beta v_{\check{\jmath}}$	(7.3.3a), (7.5.2a)	(7.5.3a), (7.3.2a)	(7.5.3a), (7.3.2a)
$\partial_\gamma v_{\check{\imath}} \partial_\alpha \partial_\beta v_{\check{\jmath}}$	(7.3.2a), (7.5.2a)	(7.5.2a), (7.3.2a)	(7.5.2a), (7.3.2a)

There are four terms to be estimated separately:

$$Z^I\big(\partial_\alpha \Psi_\beta^{\beta'} u_{\hat{\imath}} \underline{\partial}_{\beta'} u_{\hat{\jmath}}\big), \quad Z^I\big(\partial_\alpha \Psi_\beta^{\beta'} v_{\check{\imath}} \underline{\partial}_{\beta'} u_{\hat{\jmath}}\big),$$
$$Z^I\big(\partial_\alpha \Psi_\beta^{\beta'} \partial_\gamma u_{\hat{\imath}} \underline{\partial}_{\beta'} u_{\hat{\jmath}}\big), \quad Z^I\big(\partial_\alpha \Psi_\beta^{\beta'} \partial_\gamma v_{\check{\imath}} \underline{\partial}_{\beta'} u_{\hat{\jmath}}\big).$$

As before, these terms are to be estimated by the additional decay supplied by $|Z^I \Psi_\beta^{\beta'}| \leqslant C(I) t^{-1}$. We omit the details but list out the inequalities to be used for each term and each partition of $I = I_2 + I_3$:

Products	$(2, 0)$	$(1, 1)$	$(0, 2)$
$\partial_\alpha \Psi_\beta^{\beta'} u_{\hat{\imath}} \underline{\partial}_{\beta'} u_{\hat{\jmath}}$	(7.3.6c), (7.5.3a)	(7.5.5c), (7.3.3a)	(7.5.5c), (7.3.3a)
$\partial_\alpha \Psi_\beta^{\beta'} v_{\check{\imath}} \underline{\partial}_{\beta'} u_{\hat{\jmath}}$	(7.3.1e), (7.5.3a)	(7.5.1e), (7.3.3a)	(7.5.1e), (7.3.3a)
$\partial_\alpha \Psi_\beta^{\beta'} \partial_\gamma u_{\hat{\imath}} \underline{\partial}_{\beta'} u_{\hat{\jmath}}$	(7.3.3a), (7.5.3a)	(7.5.3a), (7.3.3a)	(7.5.3a), (7.3.3a)
$\partial_\alpha \Psi_\beta^{\beta'} \partial_\gamma v_{\check{\imath}} \underline{\partial}_{\beta'} u_{\hat{\jmath}}$	(7.3.2a), (7.5.3a)	(7.5.2a), (7.3.3a)	(7.5.2a), (7.3.3a)

And we conclude with (8.5.3a).

We turn our attention to the estimates for $Q_{T\hat{\imath}}$. As before the details are omitted. The inequalities we use for each term and each partition of the indices are listed:

Products	$(2, \leqslant 0)$	$(1, \leqslant 1)$	$(0, \leqslant 2)$
$s\underline{\partial}_a \underline{\partial}_\beta Z^{I_1} u_{\hat{\imath}} Z^{I_2} \partial_\gamma u_{\hat{k}}$	(7.3.4c), (7.5.3a)	(7.3.4c), (7.5.3a)	(7.4.5c), (7.3.3a)
$s\underline{\partial}_a \underline{\partial}_\beta Z^{I_1} u_{\hat{\imath}} Z^{I_2} \partial_\gamma v_{\check{k}}$	(7.3.4c), (7.5.2a)	(7.3.4c), (7.5.2a)	(7.4.5c), (7.3.2a)
$s\underline{\partial}_a \underline{\partial}_\beta Z^{I_1} u_{\hat{\imath}} Z^{I_2} u_{\hat{k}}$	(7.3.4c), (7.5.5c)	(7.3.4c), (7.5.5c)	(7.4.5c), (7.3.6c)
$s\underline{\partial}_a \underline{\partial}_\beta Z^{I_1} u_{\hat{\imath}} Z^{I_2} v_{\check{k}}$	(7.3.4c), (7.5.1e)	(7.3.4c), (7.5.1e)	(7.4.5c), (7.3.1e)

Products	$(2, \leqslant 0)$	$(1, \leqslant 1)$	$(0, \leqslant 2)$
$st^{-1} \partial_{\gamma'} Z^{I_1} u_{\hat{\jmath}} Z^{I_2} \partial_\gamma u_{\hat{k}}$	(7.2.3a), (7.5.3a)	(7.2.3a), (7.5.3a)	(7.4.4a), (7.3.3a)
$st^{-1} \partial_{\gamma'} Z^{I_1} u_{\hat{\jmath}} Z^{I_2} \partial_\gamma v_{\check{k}}$	(7.2.3a), (7.5.2a)	(7.2.3a), (7.5.2a)	(7.4.4a), (7.3.2a)
$st^{-1} \partial_{\gamma'} Z^{I_1} u_{\hat{\jmath}} Z^{I_2} u_{\hat{k}}$	(7.2.3a), (7.5.5c)	(7.2.3a), (7.5.5c)	(7.4.4a), (7.3.6c)
$st^{-1} \partial_{\gamma'} Z^{I_1} u_{\hat{\jmath}} Z^{I_2} v_{\check{k}}$	(7.2.3a), (7.5.1e)	(7.2.3a), (7.5.1e)	(7.4.4a), (7.3.1e)

The estimate of $Z^I F_{\hat\imath}$ is essentially the same. As before, we omit the details but list out the inequalities to be used:

Products	(2,0)	(1,1)	(0,2)
$\partial_\alpha u_{\hat\jmath}\partial_\beta u_{\hat k}$	(7.3.3a), (7.5.3a)	(7.5.3a), (7.3.3a)	(7.5.3a), (7.3.3a)
$\partial_\alpha v_{\check\jmath}\partial_\beta w_k$	(7.3.2a), (7.5.1a)	(7.5.2a), (7.3.1a)	(7.5.2a), (7.3.1a)
$v_{\check k}\partial_\alpha w_j$	(7.3.1e), (7.5.1a)	(7.5.1e), (7.3.1a)	(7.5.1e), (7.3.1a)
$v_{\check k}v_{\check\jmath}$	(7.3.1e), (7.5.1e)	(7.5.1e), (7.3.1e)	(7.5.1e), (7.3.1e)

The estimate of $R(Z^I u_{\hat\imath})$ is a direct result of (7.3.4c). □

Now we are ready to prove the second main result of this section.

Proposition 8.5.1. *Let $u_{\hat\imath}$ be wave components of a sufficiently regular, local-in-time solution to* (1.2.1) *and assume that* (2.4.5) *holds with $C_1\epsilon \leqslant \min\{1, \epsilon_0''\}$. The following estimates hold for $|I^\dagger| \leqslant 4$, $|I| \leqslant 3$ and $|I^\flat| \leqslant 2$:*

$$\|s^3t^{-2}\partial_t\partial_t Z^{I^\dagger}u_{\hat\imath}\|_{L^2(\mathcal{H}_s)} \leqslant CC_1\epsilon s^\delta, \tag{8.5.4a}$$

$$\|s^3t^{-2}\partial_t\partial_t Z^{I}u_{\hat\imath}\|_{L^2(\mathcal{H}_s)} \leqslant CC_1\epsilon s^{\delta/2}, \tag{8.5.4b}$$

$$\|s^3t^{-2}\partial_t\partial_t Z^{I^\flat}u_{\hat\imath}\|_{L^2(\mathcal{H}_s)} \leqslant CC_1\epsilon \tag{8.5.4c}$$

and, furthermore,

$$\|s^3t^{-2}\partial_\alpha\partial_\beta Z^{I^\dagger}u_{\hat\imath}\|_{L^2(\mathcal{H}_s)} \leqslant CC_1\epsilon s^\delta, \tag{8.5.5a}$$

$$\|s^3t^{-2}\partial_\alpha\partial_\beta Z^{I}u_{\hat\imath}\|_{L^2(\mathcal{H}_s)} \leqslant CC_1\epsilon s^{\delta/2}, \tag{8.5.5b}$$

$$\|s^3t^{-2}\partial_\alpha\partial_\beta Z^{I^\flat}u_{\hat\imath}\|_{L^2(\mathcal{H}_s)} \leqslant CC_1\epsilon. \tag{8.5.5c}$$

Proof. The proof is a combination of (8.5.1), (8.5.2), and (8.5.3) with (8.2.5). We will prove first (8.5.4c).

The proof is done by induction. We first prove (8.5.4c) with $|I^\flat| = 0$ and recall our convention that $|I| < 0$ implies $Z^I = 0$, and (8.2.5) implies (with $|I| = 0$):

$$\begin{aligned}&\big\|s^3t^{-2}\partial_t\partial_t u_{\hat\imath}\big\|_{L^2(\mathcal{H}_s)}\\ &\leqslant C\big\|sQ_{G\hat\imath}\big\|_{L^2(\mathcal{H}_s)} + C\big\|sQ_{T\hat\imath}\big\|_{L^2(\mathcal{H}_s)} + C\big\|sF_{\hat\imath}\big\|_{L^2(\mathcal{H}_s)} + C\big\|sR(u_{\hat\imath})\big\|_{L^2(\mathcal{H}_s)}.\end{aligned}$$

By the group of inequalities (8.5.3),

$$\big\|s^3t^{-2}\partial_t\partial_t u_{\hat\imath}\big\|_{L^2(\mathcal{H}_s)} \leqslant CC_1\epsilon.$$

Suppose that (8.5.4c) holds with $|I^\flat| \leqslant m$, we will prove (8.5.4c) with $|I^\flat| \leqslant m+1$. By (8.2.5),

$$\begin{aligned}
&\big\|s^3t^{-2}\partial_t\partial_t Z^{I^\flat} u_{\hat{\imath}}\big\|_{L^2(\mathcal{H}_s)}\\
&\leqslant CK \sum_{\substack{|I_2|+|I_3|\leqslant|I^\flat|\\|I_2|<|I^\flat|}} \sum_{\gamma,\hat{\jmath},i} \big\|s(|Z^{I_3}\partial_\gamma w_i| + |Z^{I_3}w_i|)\,\partial_t\partial_t Z^{I_2}u_{\hat{\jmath}}\big\|_{L^2(\mathcal{H}_s)}\\
&\quad + C\|sQ_{G\hat{\imath}}(I^\flat, w, \partial w, \partial\partial w)_{\hat{\imath}}\|_{L^2(\mathcal{H}_s)} + C\|sQ_{T\hat{\imath}}(I^\flat, w, \partial w, \partial\partial w)\|_{\hat{\imath}}\|_{L^2(\mathcal{H}_s)}\\
&\quad + C\|sR(Z^{I^\flat}u_{\hat{\imath}})\|_{L^2(\mathcal{H}_s)}
\end{aligned}$$

when $|I^\flat| \leqslant m+1 \leqslant 2$, and we apply (8.5.3):

$$\begin{aligned}
&\big\|s^3t^{-2}\partial_t\partial_t Z^{I^\flat} u_{\hat{\imath}}\big\|_{L^2(\mathcal{H}_s)}\\
&\leqslant CK \sum_{\substack{|I_2|+|I_3|\leqslant|I^\flat|\\|I_2|<|I^\flat|}} \sum_{\gamma,\hat{\jmath},i} \big\|s(|Z^{I_3}\partial_\gamma w_i| + |Z^{I_3}w_i|)\,\partial_t\partial_t Z^{I_2}u_{\hat{\jmath}}\big\|_{L^2(\mathcal{H}_s)} + CC_1\epsilon\\
&\leqslant CK \sum_{\substack{|I_2|+|I_3|\leqslant|I^\flat|\\|I_3|=1}} \sum_{\gamma,\hat{\jmath},i} \big\|s(|Z^{I_3}\partial_\gamma w_i| + |Z^{I_3}w_i|)\,\partial_t\partial_t Z^{I_2}u_{\hat{\jmath}}\big\|_{L^2(\mathcal{H}_s)}\\
&\quad + CK \sum_{\substack{|I_2|+|I_3|\leqslant|I^\flat|\\|I_3|=2}} \sum_{\gamma,\hat{\jmath},i} \big\|s(|Z^{I_3}\partial_\gamma w_i| + |Z^{I_3}w_i|)\,\partial_t\partial_t Z^{I_2}u_{\hat{\jmath}}\big\|_{L^2(\mathcal{H}_s)} + CC_1\epsilon.
\end{aligned}$$

Observe now that when $I_3 = 1$, by (7.5.3a), (7.5.1c) and (7.5.5a) we have

$$\begin{aligned}
&|Z^{I_3}\partial_\gamma u_{\hat{\imath}}| \leqslant CC_1\epsilon t^{-1/2}s^{-1}, \quad |Z^{I_3}\partial_\gamma v_{\check{\imath}}| \leqslant CC_1\epsilon t^{-3/2}s^{\delta},\\
&|Z^{I_3}u_{\hat{\imath}}| \leqslant CC_1\epsilon t^{-3/2}s, \quad |Z^{I_3}v_{\check{\jmath}}| \leqslant CC_1\epsilon t^{-3/2}s^{\delta}.
\end{aligned}$$

And by the induction assumption $\big((8.5.4\text{c})$ for $|I_2| \leqslant |I^\flat| - 1\big)$:

$$\begin{aligned}
&\sum_{\substack{|I_2|+|I_3|\leqslant|I^\flat|\\|I_3|=1}} \sum_{\gamma,\hat{\jmath},i} \big\|s(|Z^{I_3}\partial_\gamma w_i| + |Z^{I_3}w_i|)\,\partial_t\partial_t Z^{I_2}u_{\hat{\jmath}}\big\|_{L^2(\mathcal{H}_s)}\\
&\leqslant C(C_1\epsilon) \sum_{\substack{|I_2|+|I_3|\leqslant|I^\flat|\\\gamma,\hat{\jmath},i,|I_3|=1}} \big\|t^2s^{-2}(t^{-1/2}s^{-1} + t^{-3/2}s + t^{-3/2}s^{\delta})\,s^3t^{-2}\partial_t\partial_t Z^{I_2}u_{\hat{\jmath}}\big\|_{L^2(\mathcal{H}_s)}\\
&\leqslant C(C_1\epsilon) \sum_{\substack{|I_2|+|I_3|\leqslant|I^\flat|\\|I_3|=1}} \sum_{\gamma,\hat{\jmath},i} \big\|s^3t^{-2}\partial_t\partial_t Z^{I_2}u_{\hat{\jmath}}\big\|_{L^2(\mathcal{H}_s)}\\
&\leqslant C(C_1\epsilon)^2.
\end{aligned}$$

When $|I_3| = 2$, we observe that $|I_2| \leqslant 0$. By (8.4.4c), we have

$$\begin{aligned}
&\sum_{\gamma,\widehat{\jmath},i} \big\|s(|Z^{I_3}\partial_\gamma w_i| + |Z^{I_3}w_i|)\,\partial_t\partial_t Z^{I_2}u_{\widehat{\jmath}}\big\|_{L^2(\mathcal{H}_s)} \\
&\leqslant CC_1\epsilon \sum_{\gamma,\widehat{\jmath},i} \big\|s(|Z^{I_3}\partial_\gamma w_i| + |Z^{I_3}w_i|)\,t^{1/2}s^{-3}\big\|_{L^2(\mathcal{H}_s)} \\
&\leqslant CC_1\epsilon \sum_{\gamma,\widehat{\imath}} \big\|t^{1/2}s^{-2}Z^{I_3}\partial_\gamma u_{\widehat{\imath}}\big\|_{L^2(\mathcal{H}_s)} + CC_1\epsilon \sum_{\gamma,\check{\imath}} \big\|t^{1/2}s^{-2}Z^{I_3}\partial_\gamma v_{\check{\imath}}\big\|_{L^2(\mathcal{H}_s)} \\
&\quad + CC_1\epsilon \sum_{\widehat{\imath}} \big\|t^{1/2}s^{-2}Z^{I_3}u_{\widehat{\imath}}\big\|_{L^2(\mathcal{H}_s)} + CC_1\epsilon \sum_{\check{\imath}} \big\|t^{1/2}s^{-2}Z^{I_3}v_{\check{\imath}}\big\|_{L^2(\mathcal{H}_s)}
\end{aligned}$$

These four terms can be bounded by $C(C_1\epsilon)^2$ by applying (7.3.3a), (7.3.2a) and (7.3.6c). So for $|I^\flat| \leqslant m+1 \leqslant 2$, (8.5.4c) is proved. By induction, (8.5.4c) is proved for $|I^\flat| \leqslant 2$.

We turn to the proof of (8.5.4b) and observe that in (8.5.4b), the case $|I| \leqslant 2$ is already proved by (8.5.4c). We need only treat the cases $|I| = 3$.

When $|I| = 3$, we apply (8.2.5):

$$\begin{aligned}
&\big\|s^3t^{-2}\partial_t\partial_t Z^I u_{\widehat{\imath}}\big\|_{L^2(\mathcal{H}_s)} \\
&\leqslant CK \sum_{\substack{|I_2|+|I_3|\leqslant 3\\ |I_2|<3}} \sum_{\gamma,\widehat{\jmath},i} \big\|s(|Z^{I_3}\partial_\gamma w_i| + |Z^{I_3}w_i|)\,\partial_t\partial_t Z^{I_2}u_{\widehat{\jmath}}\big\|_{L^2(\mathcal{H}_s)} \\
&\quad + C\|sQ_{G\widehat{\imath}}(I,w,\partial w,\partial\partial w)_{\widehat{\imath}}\|_{L^2(\mathcal{H}_s)} + C\|Z^I F_{\widehat{\imath}}\| + C\|sZ^I u_{\widehat{\imath}}\|_{L^2(\mathcal{H}_s)}.
\end{aligned}$$

We observe that, in view of (8.5.2),

$$\begin{aligned}
&\big\|s^3t^{-2}\partial_t\partial_t Z^I u_{\widehat{\imath}}\big\|_{L^2(\mathcal{H}_s)} \\
&\leqslant CK \sum_{\substack{|I_2|+|I_3|\leqslant 3\\ |I_2|<3}} \sum_{\gamma,\widehat{\jmath},i} \big\|s(|Z^{I_3}\partial_\gamma w_i| + |Z^{I_3}w_i|)\,\partial_t\partial_t Z^{I_2}u_{\widehat{\jmath}}\big\|_{L^2(\mathcal{H}_s)} + CC_1\epsilon s^{\delta/2}.
\end{aligned}$$

We focus on

$$\begin{aligned}
&\sum_{\substack{|I_2|+|I_3|\leqslant 3\\ |I_2|<3}} \sum_{\gamma,\widehat{\jmath},i} \big\|s(|Z^{I_3}\partial_\gamma w_i| + |Z^{I_3}w_i|)\,\partial_t\partial_t Z^{I_2}u_{\widehat{\jmath}}\big\|_{L^2(\mathcal{H}_s)} \\
&\leqslant \sum_{\substack{|I_2|+|I_3|\leqslant 3\\ |I_2|=0}} \sum_{\gamma,\widehat{\jmath},\widehat{\imath}} \big\|sZ^{I_3}\partial_\gamma u_{\widehat{\imath}}\,\partial_t\partial_t Z^{I_2}u_{\widehat{\jmath}}\big\|_{L^2(\mathcal{H}_s)} \\
&\quad + \sum_{\substack{|I_2|+|I_3|\leqslant 3\\ |I_2|=1}} \sum_{\gamma,\widehat{\jmath},\check{\imath}} \big\|sZ^{I_3}\partial_\gamma v_{\check{\imath}}\,\partial_t\partial_t Z^{I_2}u_{\widehat{\jmath}}\big\|_{L^2(\mathcal{H}_s)} \\
&\quad + \sum_{\substack{|I_2|+|I_3|\leqslant 3\\ |I_2|=2}} \sum_{\gamma,\widehat{\jmath},\widehat{\imath}} \big\|sZ^{I_3}u_{\widehat{\imath}}\,\partial_t\partial_t Z^{I_2}u_{\widehat{\jmath}}\big\|_{L^2(\mathcal{H}_s)}.
\end{aligned}$$

For each sum, the possible choice of $(|I_2|,|I_3|)$ are

$$(0, \leqslant 3), \quad (1, \leqslant 2), \quad (2, \leqslant 1).$$

We list out, for each sum, the relevant inequalities for each possible choice $(|I_2|,|I_3|)$. This makes the following list:

Products	$(0,\leqslant 3)$	$(1,\leqslant 2)$	$(2,\leqslant 1)$
$s\partial_t\partial_t Z^{I_2}u_{\hat{\jmath}}\, Z^{I_3}\partial_\gamma u_{\hat{\imath}}$	(8.4.6c), (7.3.3a)	(8.5.4c), (7.5.2b)	(8.5.4c), (7.5.3a)
$s\partial_t\partial_t Z^{I_2}u_{\hat{\jmath}}\, Z^{I_3}\partial_\gamma v_{\check{\imath}}$	(8.4.6c), (7.3.2b)	(8.5.4c), (7.5.2a)	(8.5.4c), (7.5.2a)
$s\partial_t\partial_t Z^{I_2}u_{\hat{\jmath}}\, Z^{I_3}u_{\hat{\imath}}$	(8.4.6c), (7.3.6b)	(8.5.4c), (7.5.5b)	(8.5.4c), (7.5.5c)
$s\partial_t\partial_t Z^{I_2}u_{\hat{\jmath}}\, Z^{I_3}v_{\check{\imath}}$	(8.4.6c), (7.3.2b)	(8.5.4c), (7.5.2a)	(8.5.4c), (7.5.2a)

We conclude with the fact that these four terms can be bounded by $C(C_1\epsilon)^2 s^{\delta/2}$. So (8.5.4b) is proved for $|I^\dagger| = 3$.

Now we turns to the proof of (8.5.4). When $|I^\dagger| = 4$, we apply (8.5.4b) with $|I^\dagger| = 3$ and, more precisely,

$$\|s^3 t^{-2}\partial_t\partial_t Z^{I_1}u_{\hat{\imath}}\|_{L^2(\mathcal{H}_s)} \leqslant CC_1\epsilon s^{\delta/2} \leqslant CC_1\epsilon s^{\delta}. \tag{8.5.6}$$

As in previous cases, by (8.2.5) and (8.5.1),

$$\begin{aligned}
&\big\|s^3 t^{-2}\partial_t\partial_t Z^{I^\sharp}u_{\hat{\imath}}\big\|_{L^2(\mathcal{H}_s)} \\
&\leqslant CK \sum_{\substack{|I_2|+|I_3|\leqslant 4\\ |I_2|<3}} \sum_{\gamma,\hat{\jmath},i} \big\|s(|Z^{I_3}\partial_\gamma w_i| + |Z^{I_3}w_i|)\,\partial_t\partial_t Z^{I_2}u_{\hat{\jmath}}\big\|_{L^2(\mathcal{H}_s)} + C(C_1\epsilon)^2 s^\delta.
\end{aligned}$$

The first sum is also decomposed into four parts:

$$\begin{aligned}
&\sum_{\substack{|I_2|+|I_3|\leqslant 4\\ |I_2|<4}} \sum_{\gamma,\hat{\jmath},i} \big\|s(|Z^{I_3}\partial_\gamma w_i| + |Z^{I_3}w_i|)\,\partial_t\partial_t Z^{I_2}u_{\hat{\jmath}}\big\|_{L^2(\mathcal{H}_s)} \\
\leqslant\ & \sum_{\substack{|I_2|+|I_3|\leqslant 4\\ |I_2|<4}} \sum_{\gamma,\hat{\jmath},\hat{\imath}} \big\|sZ^{I_3}\partial_\gamma u_{\hat{\imath}}\,\partial_t\partial_t Z^{I_2}u_{\hat{\jmath}}\big\|_{L^2(\mathcal{H}_s)} \\
&+ \sum_{\substack{|I_2|+|I_3|\leqslant 4\\ |I_2|<4}} \sum_{\gamma,\hat{\jmath},\check{\imath}} \big\|sZ^{I_3}\partial_\gamma v_{\check{\imath}}\,\partial_t\partial_t Z^{I_2}u_{\hat{\jmath}}\big\|_{L^2(\mathcal{H}_s)} \\
&+ \sum_{\substack{|I_2|+|I_3|\leqslant 4\\ |I_2|<4}} \sum_{\gamma,\hat{\jmath},\hat{\imath}} \big\|sZ^{I_3}u_{\hat{\imath}}\,\partial_t\partial_t Z^{I_2}u_{\hat{\jmath}}\big\|_{L^2(\mathcal{H}_s)} \\
&+ \sum_{\substack{|I_2|+|I_3|\leqslant 4\\ |I_2|<4}} \sum_{\gamma,\hat{\jmath},\check{\imath}} \big\|sZ^{I_3}v_{\check{\imath}}\,\partial_t\partial_t Z^{I_2}u_{\hat{\jmath}}\big\|_{L^2(\mathcal{H}_s)}.
\end{aligned}$$

The possible choices of $(|I_2|,|I_3|)$ are

$$(3, \leqslant 1), \quad (2, \leqslant 2), \quad (1, \leqslant 3), \quad (0, \leqslant 4).$$

We list out the inequalities to be used for each term and each choice of $(|I_2|,|I_3|)$. Cf. Table 13. We conclude with (8.5.4a). On the other hand, (8.5.5) are combinations of (8.5.4) and Lemma 8.1.1. □

Table 13

Products	$(3, \leqslant 1)$	$(2, \leqslant 2)$	$(1, \leqslant 3)$	$(0, \leqslant 4)$
$s\partial_t\partial_t Z^{I_2}u_{\hat{\jmath}}\,Z^{I_3}\partial_\gamma u_{\hat{\imath}}$	(8.5.6), (7.5.3a)	(8.4.4a), (7.3.3a)	(8.4.4a), (7.3.3a)	(8.4.4c), (7.3.1a)
$s\partial_t\partial_t Z^{I_2}u_{\hat{\jmath}}\,Z^{I_3}\partial_\gamma v_{\check{\imath}}$	(8.5.6), (7.5.2a)	(8.4.4a), (7.3.2a)	(8.4.4a), (7.3.2a)	(8.4.4c), (7.3.2a)
$s\partial_t\partial_t Z^{I_2}u_{\hat{\jmath}}Z^{I_3}u_{\hat{\imath}}$	(8.5.6), (7.5.5c)	(8.4.4a), (7.3.6c)	(8.4.4a), (7.3.6c)	(8.4.4c), (7.3.6a)
$s\partial_t\partial_t Z^{I_2}u_{\hat{\jmath}}Z^{I_3}v_{\check{\imath}}$	(8.5.6), (7.5.1e)	(8.4.4a), (7.3.1c)	(8.4.4a), (7.3.1c)	(8.4.4c), (7.3.1c)

Table 14

Terms	$(3, \leqslant 0)$	$(2, \leqslant 1)$	$(1, \leqslant 2)$	$(0, \leqslant 3)$
$(s/t)^2 Z^{I_1}\partial_t u_{\hat{\imath}} Z^{I_2}\partial_t u_{\hat{\jmath}}$	(7.3.3a), (7.5.3a)	(7.3.3a), (7.5.3a)	(7.5.3a), (7.3.3a)	(7.5.3a), (7.3.3a)
$Z^{I_1}\underline{\partial}_a u_{\hat{\imath}} Z^{I_2}\underline{\partial}_\beta u_{\hat{\jmath}}$	(7.3.3b), (7.5.3a)	(7.3.3b), (7.5.3a)	(7.5.3b), (7.3.3a)	(7.5.3b), (7.3.3a)
$Z^{I_1}\underline{\partial}_\beta u_{\hat{\imath}} Z^{I_2}\underline{\partial}_a u_{\hat{\jmath}}$	(7.3.3a), (7.5.3b)	(7.3.3a), (7.5.3b)	(7.5.3a), (7.3.3b)	(7.5.3a), (7.3.3b)

Now, we derive the complete L^2 estimates of the second-order derivatives.

Lemma 8.5.4. *Under the assumption* (2.4.5), *the following* L^2 *estimates hold for* $|I^\dagger| \leqslant 4, |I| \leqslant 3|$ *and* $I^\flat| \leqslant 2$*:*

$$\|s^3t^{-2}\partial_\alpha\partial_\beta Z^{I^\dagger} u_{\hat{\imath}}\|_{L^2(\mathcal{H}_s)} + \|s^3t^{-2}Z^{I^\dagger}\partial_\alpha\partial_\beta u_{\hat{\imath}}\|_{L^2(\mathcal{H}_s)} \leqslant CC_1\epsilon s^\delta, \quad (8.5.7a)$$

$$\|s^3t^{-2}\partial_\alpha\partial_\beta Z^{I} u_{\hat{\imath}}\|_{L^2(\mathcal{H}_s)} + \|s^3t^{-2}Z^{I}\partial_\alpha\partial_\beta u_{\hat{\imath}}\|_{L^2(\mathcal{H}_s)} \leqslant CC_1\epsilon s^{\delta/2}, \quad (8.5.7b)$$

$$\|s^3t^{-2}\partial_\alpha\partial_\beta Z^{I^\flat} u_{\hat{\imath}}\|_{L^2(\mathcal{H}_s)} + \|s^3t^{-2}Z^{I^\flat}\partial_\alpha\partial_\beta u_{\hat{\imath}}\|_{L^2(\mathcal{H}_s)} \leqslant CC_1\epsilon. \quad (8.5.7c)$$

Proof. These inequalities are based on (3.3.3) and (8.5.5). □

Chapter 9

Null forms and decay in time

9.1 Bounds that are independent of second-order estimates

In this chapter, we are going to estimate the terms

$$\mathcal{T}(\partial u_{\hat{\imath}}, \partial u_{\hat{\jmath}}) := T^{\alpha\beta}\partial_\alpha u_{\hat{\imath}} \partial_\beta u_{\hat{\jmath}}, \quad [Z^I, A^{\alpha\beta\gamma}\partial_\gamma u_{\hat{\imath}} \partial_\alpha \partial_\beta] u_{\hat{\jmath}}$$

where $T^{\alpha\beta}$ and $A^{\alpha\beta\gamma}$ are null quadratic forms. Concerning the term $[Z^I, A^{\alpha\beta\gamma}\partial_\gamma u_{\hat{\imath}} \partial_\alpha \partial_\beta] u_{\hat{\jmath}}$, we can apply (8.5.4c) and (8.4.4c) for better decay rates, but the following rates will be sufficient for our main result in this monograph.

Lemma 9.1.1. *By reyling on the energy assumption* (2.4.5e), *the following estimates hold for all* $|I| \leqslant 3$ *and* $|J| \leqslant 1$*:*

$$\big\| Z^I \mathcal{T}(\partial u_{\hat{\imath}}, \partial u_{\hat{\jmath}}) \big\|_{L^2(\mathcal{H}_s)} \leqslant C(C_1\epsilon)^2 s^{-3/2}, \tag{9.1.1a}$$

$$\big\| [Z^I, A^{\alpha\beta\gamma}\partial_\gamma u_{\hat{\imath}} \partial_\alpha \partial_\beta] u_{\hat{\jmath}} \big\|_{L^2(\mathcal{H}_s)} \leqslant C(C_1\epsilon)^2 s^{-3/2+\delta}, \tag{9.1.1b}$$

$$\sup_{\mathcal{H}_s} \big(t^2 s \mathcal{T}(\partial u_{\hat{\imath}}, \partial u_{\hat{\jmath}}) \big) \leqslant C(C_1\epsilon)^2. \tag{9.1.2}$$

Proof. The proof of (9.1.1a) is a combination of Proposition 4.1.2 with (7.5.3) and (7.3.3). By recalling the decomposition of $\mathcal{T}$ presented in Proposition 4.1.2, we have the list in Table 14. This completes the argument.

The estimate on the term $[Z^I, A^{\alpha\beta\gamma}\partial_\gamma u_{\hat{\imath}} \partial_\alpha \partial_\beta] u_{\hat{\jmath}}$ is also proved by using the inequalities presented in the following list. Observe that by Proposition 4.3.2, some partition of the index do not exist. We have:

Terms	$(3, \leqslant 0)$	$(2, \leqslant 1)$	$(1, \leqslant 2)$
$(s/t)^2 Z^{I_1}\partial_t u_{\hat{k}} Z^{I_2}\partial_t\partial_t u_{\hat{\jmath}}$	(7.3.3a), (7.5.3a)	(7.3.3a), (7.5.1a)	(7.5.3a), (7.3.3a)
$Z^{I_1}\underline{\partial}_a u_{\hat{k}} Z^{I_2}\underline{\partial}_\beta\underline{\partial}_\gamma u_{\hat{\jmath}}$	(7.3.3b), (7.5.3a)	(7.3.3b), (7.5.1a)	(7.5.3b), (7.3.3a)
$Z^{I_1}\underline{\partial}_\alpha u_{\hat{k}} Z^{I_2}\underline{\partial}_b\underline{\partial}_\gamma u_{\hat{\jmath}}$	(7.3.3a), (7.5.4c)	(7.3.3a), (7.5.4a)	(7.5.3a), (7.3.5c)
$Z^{I_1}\underline{\partial}_\alpha u_{\hat{k}} Z^{I_2}\underline{\partial}_\beta\underline{\partial}_c u_{\hat{\jmath}}$	(7.3.3a), (7.5.4c)	(7.3.3a), (7.5.4a)	(7.5.3a), (7.3.5c)
$t^{-1} Z^{I_1}\partial_\alpha u_{\hat{k}}\underline{\partial}_\beta u_{\hat{\jmath}}$	(7.3.3a), (7.5.3a)	(7.3.3a), (7.5.3a)	(7.5.3a), (7.3.3a)

Finally, we derive the L^∞ estimates of $\mathcal{T}$ by the following list:

Terms	$(1, \leqslant 0)$	$(0, \leqslant 1)$	Decay rate
$(s/t)^2 Z^{J_1}\partial_t u_{\hat\imath} Z^{J_2}\partial_t u_{\hat\jmath}$	(7.5.3a), (7.5.3a)	(7.5.3a), (7.5.3a)	t^{-3}
$Z^{J_1}\underline{\partial}_a u_{\hat\imath} Z^{J_2}\underline{\partial}_\beta u_{\hat\jmath}$	(7.5.3b), (7.5.3a)	(7.5.3b), (7.5.3a)	$t^{-2}s^{-1}$
$Z^{J_1}\underline{\partial}_\beta u_{\hat\imath} Z^{J_2}\underline{\partial}_a u_{\hat\jmath}$	(7.5.3a), (7.5.3b)	(7.5.3a), (7.5.3b)	$t^{-2}s^{-1}$

□

9.2 Bounds that depend on second-order estimates

In this section we estimate the terms $[Z^I, B^{\alpha\beta}u_{\hat\imath}\partial_\alpha\partial_\beta]u_{\hat\jmath}$ by essentially relying on the second-order estimates (8.4.6).

Lemma 9.2.1. *By relying on the energy assumption* (2.4.5e), *the following estimates hold for* $|I| \leqslant 3$*:*

$$\big\|[Z^I, B^{\alpha\beta}u_{\hat\imath}\partial_\alpha\partial_\beta]u_{\hat\jmath}\big\|_{L^2(\mathcal{H}_s)} \leqslant C(C_1\epsilon)^2 s^{-3/2+\delta}. \tag{9.2.1a}$$

Proof. Recall the structure of $[Z^I, B^{\alpha\beta}u_{\hat\imath}\partial_\alpha\partial_\beta]u_{\hat\jmath}$ presented in (4.2.1b). It is a linear combinations of several terms and we need to control each term in each possible partition of the index $|I_1| + |I_2| \leqslant |I|, |I_2| < |I|$. As before, we list out the inequalities we use for each term and each partition of the indices:

terms	$(3, \leqslant 0)$	$(2, \leqslant 1)$	$(1, \leqslant 2)$
$(s/t)^2 Z^{I_1}u_{\hat k} Z^{I_2}\partial_t\partial_t u_{\hat\jmath}$	(7.3.6c), (8.4.4c)	(7.3.6c), (8.4.4a)	(7.3.6c), (8.4.4a)
$Z^{I_1}u_{\hat k} Z^{I_2}\underline{\partial}_a\underline{\partial}_\beta u_{\hat\jmath}$	(7.3.6c), (7.5.4c)	(7.3.6c), (7.5.4a)	(7.3.6c), (7.5.4a)
$Z^{I_1}u_{\hat k} Z^{I_2}\underline{\partial}_\alpha\underline{\partial}_b u_{\hat\jmath}$	(7.3.6c), (7.5.4c)	(7.3.6c), (7.5.4a)	(7.3.6c), (7.5.4a)
$t^{-1}Z^{I_1}u_{\hat k} Z^{I_1}\partial_\alpha u_{\hat\jmath}$	(7.3.6c), (7.5.3a)	(7.3.6c), (7.5.3a)	(7.5.5c), (7.3.3a)

□

9.3 Decay estimates

We are now in a position to prove (2.4.7a), (2.4.7b), (2.4.8a), and (2.4.9a). The proof of (2.4.7a) requires only the L^∞ estimate established in Chapter 7.

Lemma 9.3.1. *Let* $\{w_i\}$ *be the local-in-time solution of* (1.2.1) *and suppose that the energy assumption* (2.4.5b) *and* (2.4.5e) *hold. Then there exists* *a constant* $\kappa_1 > 0$ *(depending upon the constants* $A_i^{j\alpha\beta\gamma k}$ *and* $B_i^{j\alpha\beta k}$*) such that, if* $C_1\epsilon$ *is sufficiently small,*

$$\kappa_1^{-2}\sum_i E_{m,c_i}(s, Z^I w_i) \leqslant \sum_i E_{G,c_i}(s, Z^I w_i) \leqslant \kappa_1^2 \sum_i E_{m,c_i}(s, Z^I w_i).$$

Proof. Note that

$$\begin{aligned}\sum_{i,j,\alpha,\beta} |G_i^{j\alpha\beta}| &\leqslant CK\sum_i \big(|\partial w_i| + |w_i|\big)\\ &\leqslant CK \sum_{\widehat{\imath},\check{\imath},\alpha} \big(|\partial_\alpha v_{\check{\imath}}| + |\partial_\beta u_{\widehat{\imath}}| + |v_{\check{\imath}}| + |u_{\widehat{\imath}}|\big).\end{aligned}$$

Applying (7.5.2a), (7.5.3a), (7.5.1e), and (7.5.5c), and recalling $0 < \delta < 1/6$, we have

$$\sum_{i,j,\alpha,\beta} |G_i^{j\alpha\beta}| \leqslant CK(t^{-3/2}s + t^{-1/2}s^{-1}).$$

$$\begin{aligned}&\sum_i \big|E_{G,c_i}(s, w_i) - E_{m,c_i}(s, w_i)\big|\\ &= \left|2\int_{\mathcal{H}_s} (\partial_t w_i \partial_\beta w_j G_i^{j\alpha\beta})_{0\leqslant\alpha\leqslant 1}(1, -x/t)dx - \int_{\mathcal{H}_s} (\partial_\alpha w_i \partial_\beta w_j G_i^{j\alpha\beta})dx\right|\\ &\leqslant 2\int_{\mathcal{H}_s}\Big(\sum_{i,j,\alpha,\beta}|G_i^{j\alpha\beta}|\Big)\Big(\sum_{\alpha,k}|\partial_\alpha w_k|^2\Big)dx\\ &\leqslant 2CK\int_{\mathcal{H}_s}\sum_i\big(|\partial w_i| + |w_i|\big)\Big(\sum_{\alpha,k}|\partial_\alpha w_k|^2\Big)dx\\ &\leqslant 2CKC_1\epsilon\int_{\mathcal{H}_s}(t^{-1/2}s^{-1} + t^{-3/2}s)(t/s)^2\Big(\sum_{\alpha,k}|(s/t)\partial_\alpha w_k|^2\Big)dx\\ &\leqslant CKC_1\epsilon\sum_i E_m(s, w_i) \leqslant CKC_1\epsilon\sum_i E_{m,c_i}(s, w_i),\end{aligned}$$

where the relation $t \leqslant Cs^2$ in $\mathcal{K}$ is taken into consideration. Here we take $CKC_1\epsilon \leqslant 2/3$ with C a universal constant, then the lemma is proved by fixing $\kappa_1 = \sqrt{3}$. □

The proof of (2.4.7b) will be related to the energy estimate (7.2.1a).

Lemma 9.3.2. *By relying on* (2.4.5a), (2.4.5b), *and* (2.4.5e), *for any* $|I^\sharp| \leqslant 5$ *the following estimate holds:*

$$\begin{aligned}&\left|\int_{\mathcal{H}_s}\frac{s}{t}\Big((\partial_\alpha G_i^{j\alpha\beta})\partial_t Z^{I^\sharp} w_i \partial_\beta Z^{I^\sharp} w_j - \frac{1}{2}(\partial_t G_i^{j\alpha\beta})\partial_\alpha Z^{I^\sharp} w_i \partial_\beta Z^{I^\sharp} w_j\Big)dx\right|\\ &\leqslant C(C_1\epsilon)^2 s^{-1+\delta} E_m(s, Z^{I^\sharp} w_i)^{1/2} \leqslant C(C_1\epsilon)^2 s^{-1+\delta} E_{m,\sigma}(s, Z^{I^\sharp} w_i)^{1/2}.\end{aligned} \tag{9.3.1}$$

Proof. By (7.5.1a) and (7.5.2a), we note that

$$\begin{aligned}
\left|\partial_\alpha G_i^{j\alpha\beta}\right| \leqslant & C\sum_j |\partial_\alpha w_j| + C\sum_{j,\beta}|\partial_\alpha\partial_\beta w_j| \\
\leqslant & C\sum_{\check{\jmath},\alpha,\beta}\left(|\partial_\alpha v_{\check{\jmath}}| + |\partial_\alpha\partial_\beta v_{\check{\jmath}}|\right) + C\sum_{\hat{\jmath},\alpha,\beta}\left(|\partial_\alpha u_{\hat{\jmath}}| + |\partial_\alpha\partial_\beta u_{\hat{\jmath}}|\right) \\
\leqslant & CC_1\epsilon t^{-3/2}s^\delta + CC_1\epsilon t^{-1/2}s^{-1+\delta}.
\end{aligned}$$

By substituting this result into the expression, the first term in the left-hand side of (9.3.1) is bounded as follows:

$$\begin{aligned}
&\left\|(s/t)\left(\partial_\alpha G_i^{j\alpha\beta}\right)\partial_t Z^{I^\sharp}w_i\partial_\beta Z^{I^\sharp}w_j\right\|_{L^1(\mathcal{H}_s)} \\
&= \left\|\left((t/s)\partial_\alpha G_i^{j\alpha\beta}\right)(s/t)\partial_t Z^{I^\sharp}w_i(s/t)\partial_\beta Z^{I^\sharp}w_j\right\|_{L^1(\mathcal{H}_s)} \\
&\leqslant \sum_{j,\beta}\left\|C(C_1\epsilon)(t^{1/2}s^{-2} + t^{-1/2}s^{-1+\delta})\,(s/t)\partial_t Z^{I^\sharp}w_i(s/t)\partial_\beta Z^{I^\sharp}w_j\right\|_{L^1(\mathcal{H}_s)} \\
&\leqslant CC_1\epsilon s^{-1}\sum_{j,\beta}\left\|(s/t)\partial_t Z^{I^\sharp}w_i(s/t)\partial_\beta Z^{I^\sharp}w_j\right\|_{L^1(\mathcal{H}_s)} \\
&\leqslant CC_1\epsilon s^{-1}\sum_{j,\beta}\left\|(s/t)\partial_\beta Z^{I^\sharp}w_j\right\|_{L^2(\mathcal{H}_s)}\left\|(s/t)\partial_t Z^{I^\sharp}w_i\right\|_{L^2(\mathcal{H}_s)}.
\end{aligned}$$

We apply (7.2.1a) and find

$$\begin{aligned}
&\left\|(s/t)\left(\partial_\alpha G_i^{j\alpha\beta}\right)\partial_t Z^{I^\sharp}w_i\partial_\beta Z^{I^\sharp}w_j\right\|_{L^1(\mathcal{H}_s)} \\
&\leqslant C(C_1\epsilon)^2 s^{-1+\delta}\left(\int_{\mathcal{H}_s}\left|(s/t)\partial_t Z^{I^\sharp}w_i\right|^2 dx\right)^{1/2} \\
&\leqslant C(C_1\epsilon)^2 s^{-1+\delta}E_m(s, Z^{I^\sharp}w_i)^{1/2}.
\end{aligned}$$

The second term is estimated in the same way and we omit the details. □

The proof of (2.4.8a) is similar to that of (2.4.7b).

Lemma 9.3.3. *By relying on* (2.4.5c), (2.4.5d), *and* (2.4.5e), *the following estimate holds:*

$$\begin{aligned}
&\left|\int_{\mathcal{H}_s}\frac{s}{t}\left(\left(\partial_\alpha G_i^{j\alpha\beta}\right)\partial_t Z^{I^\dagger}w_i\partial_\beta Z^{I^\dagger}w_j - \frac{1}{2}\left(\partial_t G_i^{j\alpha\beta}\right)\partial_\alpha Z^{I^\dagger}w_i\partial_\beta Z^{I^\dagger}w_j\right)dx\right| \\
\leqslant & C(C_1\epsilon)^2 s^{-1+\delta/2}E_m(s, Z^{I^\dagger}w_i)^{1/2} \leqslant C(C_1\epsilon)^2 s^{-1+\delta/2}E_{m,\sigma}(s, Z^{I^\dagger}w_i)^{1/2}.
\end{aligned} \tag{9.3.2}$$

Proof. As in the proof of Lemma 9.3.2, we can apply (7.2.1b) together with (7.5.1b) and (7.5.2b):

$$
\begin{aligned}
&\big\|(s/t)\big(\partial_\alpha G_{\imath}^{\jmath\alpha\beta}\big)\partial_t Z^{I^\sharp} w_i \partial_\beta Z^{I^\sharp} w_j\big\|_{L^1(\mathcal{H}_s)} \\
&\leqslant C(C_1\epsilon)^2 s^{-1+\delta/2}\bigg(\int_{\mathcal{H}_s}\big|(s/t)\partial_t Z^{I^\sharp} w_i\big|^2 dx\bigg)^{1/2} \\
&\leqslant C(C_1\epsilon)^2 s^{-1+\delta/2} E_m(s, Z^{I^\sharp} w_i)^{1/2}.
\end{aligned}
$$

The second term is estimated in the same way and we omit the details. □

The proof of (2.4.9a) is quite similar, but the null structure will be taken into consideration for the sharp decay rate $s^{-3/2+2\delta}$. This is one of the only two places where the null structure is taken into account.

Lemma 9.3.4. *Suppose* (2.4.5b) *and* (2.4.5e) *hold, then for any* $|I| \leqslant 3$ *the following estimates are true:*

$$
\begin{aligned}
&\bigg|\int_{\mathcal{H}_s}\frac{s}{t}\Big(\big(\partial_\alpha G_{\hat{\imath}}^{\hat{\jmath}\alpha\beta}\big)\partial_t Z^I u_{\hat{\imath}}\partial_\beta Z^I u_{\hat{\jmath}} - \frac{1}{2}\big(\partial_t G_{\hat{\imath}}^{\hat{\jmath}\alpha\beta}\big)\partial_\alpha Z^I u_{\hat{\imath}}\partial_\beta Z^I u_{\hat{\jmath}}\Big)dx\bigg| \\
&\leqslant C(C_1\epsilon)^2 s^{-3/2+\delta} E_m(s, Z^I u_{\hat{\imath}})^{1/2}.
\end{aligned}
\tag{9.3.3}
$$

Proof. We decompose the term $G_{\hat{\imath}}^{\hat{\jmath}\alpha\beta}$ as follows:

$$
\begin{aligned}
&(s/t)\partial_\alpha G_{\hat{\imath}}^{\hat{\jmath}\alpha\beta}\partial_\beta Z^I u_{\hat{\jmath}}\partial_t Z^I u_{\hat{\imath}} \\
&= (s/t)\big(A_{\hat{\imath}}^{\hat{\jmath}\alpha\beta\gamma\check{k}}\partial_\alpha\partial_\gamma v_{\check{k}} + B_{\hat{\imath}}^{\hat{\jmath}\alpha\beta\check{k}}\partial_\alpha v_{\check{k}}\big)\partial_\beta Z^I u_{\hat{\jmath}}\partial_t Z^I u_{\hat{\imath}} \\
&\quad + (s/t)\big(A_{\hat{\imath}}^{\hat{\jmath}\alpha\beta\gamma\hat{k}}\partial_\alpha\partial_\gamma u_{\hat{k}} + B_{\hat{\imath}}^{\hat{\jmath}\alpha\beta\hat{k}}\partial_\alpha u_{\hat{k}}\big)\partial_\beta Z^I u_{\hat{\jmath}}\partial_t Z^I u_{\hat{\imath}} \\
&:= R_1 + R_2.
\end{aligned}
$$

The estimate of R_1 is direct, and we simply apply the inequalities (7.5.1e):

$$
\begin{aligned}
&\|R_1\|_{L^1(\mathcal{H}_s)} = \big\|(s/t)\big(A_{\hat{\imath}}^{\hat{\jmath}\alpha\beta\gamma\check{k}}\partial_\alpha\partial_\gamma v_{\check{k}} + B_{\hat{\imath}}^{\hat{\jmath}\alpha\beta\check{k}}\partial_\alpha v_{\check{k}}\big)\,(\partial_\beta Z^I u_{\hat{\jmath}}\partial_t Z^I u_{\hat{\imath}})\big\|_{L^1(\mathcal{H}_s)} \\
&\leqslant CC_1\epsilon\,\|(t^{-3/2}s^\delta(t/s))\,(s/t)\partial_\beta Z^I u_{\hat{\jmath}}\,(s/t)\partial_t Z^I u_{\hat{\imath}}\|_{L^1(\mathcal{H}_s)} \\
&\leqslant CC_1\epsilon\,\|(t^{-1/2}s^{-1+\delta}(s/t)\partial_\beta Z^I u_{\hat{\jmath}}\|_{L^2(\mathcal{H}_s)}\,\|(s/t)\partial_t Z^I u_{\hat{\imath}}\|_{L^2(\mathcal{H}_s)} \\
&\leqslant CC_1\epsilon s^{3/2+\delta}\|(s/t)\partial_\beta Z^I u_{\hat{\jmath}}\|_{L^2(\mathcal{H}_s)}\,E_m(s, Z^I u_{\hat{\imath}})^{1/2}.
\end{aligned}
$$

In view of (7.2.3a), we get

$$
\|R_1\|_{L^1(\mathcal{H}_s)} \leqslant C(C_1\epsilon)^2 s^{-3/2+\delta} E_m(s, Z^I u_{\hat{\imath}})^{1/2}. \tag{9.3.4}
$$

The estimates on R_2 is more involved: we have the following decomposition:

$$
\begin{aligned}
&(s/t)\big(A_{\widehat{\imath}}^{\widehat{\jmath}\alpha\beta\gamma\widehat{k}}\partial_\alpha\partial_\gamma u_{\widehat{k}} + B_{\widehat{\imath}}^{\widehat{\jmath}\alpha\beta\widehat{k}}\partial_\alpha u_{\widehat{k}}\big)\partial_\beta Z^I u_{\widehat{\jmath}}\\
=&(s/t)A_{\widehat{\imath}}^{\widehat{\jmath}\alpha\beta\gamma\widehat{k}}\partial_\alpha\partial_\gamma u_{\widehat{k}}\partial_\beta Z^I u_{\widehat{\jmath}} + (s/t)B_{\widehat{\imath}}^{\widehat{\jmath}\alpha\beta\widehat{k}}\partial_\alpha u_{\widehat{k}}\partial_\beta Z^I u_{\widehat{\jmath}}\\
=:&(s/t)\mathcal{A}_{\widehat{\imath}}^{\widehat{\jmath}\widehat{k}}(\partial Z^I u_{\widehat{\jmath}}, \partial\partial u_{\widehat{k}}) + (s/t)\mathcal{T}_{\widehat{\imath}}^{\widehat{\jmath}\widehat{k}}(\partial u_{\widehat{k}}, \partial Z^I u_{\widehat{\jmath}})
\end{aligned}
$$

Recall that $\mathcal{A}_{\widehat{\imath}}^{\widehat{\jmath}\widehat{k}}$ and $\mathcal{T}_{\widehat{\imath}}^{\widehat{\jmath}\widehat{k}}$ are null forms.

Let us consider first the terms of $\mathcal{A}_{\widehat{\imath}}^{\widehat{\jmath}\widehat{k}}$. By Proposition 4.3.1 (with $I = 0$), we have

$$
\begin{aligned}
&\mathcal{A}_{\widehat{\imath}}^{\widehat{\jmath}\widehat{k}}(\partial Z^I u_{\widehat{\jmath}}, \partial\partial u_{\widehat{k}})\\
&\leqslant CK(s/t)^2|\partial_t\partial_t u_{\widehat{k}}|\,|\partial_t Z^I u_{\widehat{\jmath}}| + CK\Omega_1(0, Z^I u_{\widehat{\jmath}}, u_{\widehat{k}}) + CKt^{-1}\Omega_2(0, Z^I u_{\widehat{\jmath}}, u_{\widehat{k}})\\
&\leqslant CK(s/t)^2|\partial_t\partial_t u_{\widehat{k}}|\,|\partial_t Z^I u_{\widehat{\jmath}}|\\
&\quad + CK\sum_{a,\beta,\gamma}|\underline{\partial}_a\underline{\partial}_\beta u_{\widehat{k}}|\,|\underline{\partial}_\gamma Z^I u_{\widehat{\jmath}}| + CK\sum_{\alpha,b,\gamma}|\underline{\partial}_\alpha\underline{\partial}_b u_{\widehat{k}}|\,|\underline{\partial}_\gamma Z^I u_{\widehat{\jmath}}|\\
&\quad + CK\sum_{\alpha,\beta,c}|\underline{\partial}_\alpha\underline{\partial}_\beta u_{\widehat{k}}|\,|\underline{\partial}_c Z^I u_{\widehat{\jmath}}| + CKt^{-1}\sum_{\alpha,\beta}|\partial_\alpha Z^I u_{\widehat{\jmath}}|\,|\partial_\beta u_{\widehat{k}}|.
\end{aligned}
$$

Each term is estimated as follows:

$$
\begin{aligned}
&\|(s/t)^2\partial_t\partial_t u_{\widehat{k}}\,\partial_t Z^I u_{\widehat{\jmath}}\|_{L^2(\mathcal{H}_s)} = \|(s/t)\partial_t\partial_t u_{\widehat{k}}\,(s/t)\partial_t Z^I u_{\widehat{\jmath}}\|_{L^2(\mathcal{H}_s)}\\
&\leqslant CC_1\epsilon\|t^{-3/2}\,(s/t)\partial_t Z^I u_{\widehat{\jmath}}\|_{L^2(\mathcal{H}_s)}\\
&\leqslant CC_1\epsilon s^{-3/2}\|(s/t)\partial_t Z^I u_{\widehat{\jmath}}\|_{L^2(\mathcal{H}_s)} \leqslant C(C_1\epsilon)^2 s^{-3/2},
\end{aligned}
$$

where we applied (7.5.3a) and (7.2.3a). We have also

$$
\begin{aligned}
&\|\underline{\partial}_a\underline{\partial}_\beta u_{\widehat{k}}\,\underline{\partial}_\gamma Z^I u_{\widehat{\jmath}}\|_{L^2(\mathcal{H}_s)} = \|(t/s)\underline{\partial}_a\underline{\partial}_\beta u_{\widehat{k}}\,(s/t)\partial_t Z^I u_{\widehat{\jmath}}\|_{L^2(\mathcal{H}_s)}\\
&\leqslant CC_1\epsilon\|(t/s)t^{-3/2}s^{-1}\,(s/t)\partial_t Z^I u_{\widehat{\jmath}}\|_{L^2(\mathcal{H}_s)}\\
&\leqslant CC_1\epsilon s^{-3/2}\|(s/t)\partial_t Z^I u_{\widehat{\jmath}}\|_{L^2(\mathcal{H}_s)} \leqslant C(C_1\epsilon)^2 s^{-3/2}.
\end{aligned}
$$

The term $\|\underline{\partial}_\alpha\underline{\partial}_b u_{\widehat{k}}\,\underline{\partial}_\gamma Z^I u_{\widehat{\jmath}}\|_{L^2(\mathcal{H}_s)}$ is estimated in the same manner and we omit the details.

$$
\|\underline{\partial}_\alpha\underline{\partial}_\beta u_{\widehat{k}}\,\underline{\partial}_c Z^I u_{\widehat{\jmath}}\|_{L^2(\mathcal{H}_s)} \leqslant \|\underline{\partial}_\alpha\underline{\partial}_\beta u_{\widehat{k}}\|_{L^\infty(\mathcal{H}_s)}\|\underline{\partial}_c Z^I u_{\widehat{\jmath}}\|_{L^2(\mathcal{H}_s)} \leqslant C(C_1\epsilon)^2 s^{-3/2}
$$

where (7.5.3a) and (7.2.3b) are used.

$$
\begin{aligned}
\|t^{-1}\partial_\alpha Z^I u_{\widehat{\jmath}}\,\partial_\beta u_{\widehat{k}}\|_{L^2(\mathcal{H}_s)} &\leqslant \|t^{-1}(t/s)\partial_\beta u_{\widehat{k}}\|_{L^\infty(\mathcal{H}_s)}\|(s/t)\partial_\alpha Z^I u_{\widehat{\jmath}}\|_{L^2(\mathcal{H}_s)}\\
&\leqslant C(C_1\epsilon)^2 s^{-5/2}.
\end{aligned}
$$

We conclude with

$$
\big\|(s/t)\mathcal{A}_{\widehat{\imath}}^{\widehat{\jmath}\widehat{k}}(\partial Z^I u_{\widehat{\imath}}, \partial\partial u_{\widehat{\jmath}})\big\|_{L^2(\mathcal{H}_s)} \leqslant CK(C_1\epsilon)^2 s^{-3/2}
$$

and

$$\begin{aligned}
&\big\|(s/t)A_{\hat\imath}^{\hat\jmath\hat k}(\partial Z^I u_{\hat\imath}, \partial\partial u_{\hat\jmath})\partial_t Z^I u_{\hat\imath}\big\|_{L^1(\mathcal{H}_s)} \\
&\leqslant \|A_{\hat\imath}^{\hat\jmath\hat k}(\partial Z^I u_{\hat\imath}, \partial\partial u_{\hat\jmath})\|_{L^2(\mathcal{H})}\|(s/t)\partial_t Z^I u_{\hat\imath}\|_{L^2(\mathcal{H}_s)} \\
&\leqslant C(C_1\epsilon)^2 s^{-3/2} E_m(s, Z^I u_{\hat\imath})^{1/2}.
\end{aligned} \tag{9.3.5}$$

Next, we consider the terms of $\mathcal{T}_{\hat\imath}^{\hat\jmath\hat k}$. By (4.2.1a), $\mathcal{T}_{\hat\imath}^{\hat\jmath\hat k}(\partial Z^I u_{\hat k}, \partial u_{\hat\jmath})$ can be bounded by a linear combination of the terms presented in Proposition 4.1.2. All the estimates are based on the inequalities (7.5.3a), (7.5.3b), (7.2.1a), and (7.2.1c).

$$\begin{aligned}
\big\|(s/t)^2\partial_t Z^I u_{\hat\jmath}\partial_t u_{\hat k}\big\|_{L^2(\mathcal{H}_s)} &\leqslant \big\|(s/t)\partial_t u_{\hat k}\big\|_{L^\infty(\mathcal{H}_s)}\big\|(s/t)\partial_t Z^I u_{\hat\jmath}\big\|_{L^2(\mathcal{H}_s)} \\
&\leqslant C(C_1\epsilon)^2 s^{-3/2},
\end{aligned}$$

$$\begin{aligned}
\big\|\underline{\partial}_a u_{\hat k}\underline{\partial}_\beta Z^I u_{\hat\jmath}\big\|_{L^2(\mathcal{H}_s)} &\leqslant \big\|(t/s)\underline{\partial}_a u_{\hat k}\big\|_{L^\infty(\mathcal{H}_s)}\big\|(s/t)\underline{\partial}_\beta Z^I u_{\hat\jmath}\big\|_{L^2(\mathcal{H}_s)} \\
&\leqslant C(C_1\epsilon)^2 s^{-3/2},
\end{aligned}$$

$$\begin{aligned}
\big\|\underline{\partial}_\alpha u_{\hat k}\underline{\partial}_b Z^I u_{\hat\jmath}\big\|_{L^2(\mathcal{H}_s)} &\leqslant \big\|\underline{\partial}_\alpha u_{\hat k}\big\|_{L^\infty(\mathcal{H}_s)}\big\|\underline{\partial}_b Z^I u_{\hat\jmath}\big\|_{L^2(\mathcal{H}_s)} \\
&\leqslant C(C_1\epsilon)^2 s^{-3/2}.
\end{aligned}$$

We conclude with

$$\begin{aligned}
\big\|\mathcal{T}(\partial u_{\hat k}, \partial Z^I u_{\hat\jmath})\partial_t Z^I u_{\hat\imath}\big\|_{L^1(\mathcal{H}_s)} &\leqslant \big\|\mathcal{T}(\partial u_{\hat k}, \partial Z^I u_{\hat\jmath})\big\|_{L^2(\mathcal{H}_s)}\big\|(s/t)\partial_t Z^I u_{\hat\imath}\big\|_{L^2(\mathcal{H}_s)} \\
&\leqslant C(C_1\epsilon)^2 s^{-3/2} E_m(s, Z^I u_{\hat\imath})^{1/2}.
\end{aligned} \tag{9.3.6}$$

Combining (9.3.4), (9.3.5), and (9.3.6), we conclude that

$$\big\|\partial_\alpha\big(G_{\hat\imath}^{\hat\jmath\alpha\beta}(w, \partial w)\big)\partial_t Z^I u_{\hat\imath}\big\|_{L^1(\mathcal{H}_s)} \leqslant C(C_1\epsilon)^2 s^{-3/2+\delta} E_m(s, Z^I u_{\hat\imath})^{1/2}.$$

By a similar argument, the term $\big\|\partial_t\big(G_{\hat\imath}^{\hat\jmath\alpha\beta}(w, \partial w)\big)\partial_\alpha Z^I u_{\hat\imath}\big\|_{L^1(\mathcal{H}_s)}$ is also bounded by $C(C_1\epsilon)^2 s^{-3/2+\delta}$, and we omit the details. □

Chapter 10

L^2 estimates of the interaction terms

10.1 L^2 estimates on higher-order interaction terms

In this chapter, we establish three groups of energy estimates mentioned in the proof of Proposition 2.4.3, that is, (2.4.7c), (2.4.8b), and (2.4.9b) which are derived under the assumption (2.4.5). We emphasize that three groups of inequalities correspond to different decreasing rates in time. The proof of (2.4.7c) and (2.4.8b) is much easier than that of (2.4.9b), since, roughly speaking, it does not require the null structure, while the decreasing rate in (2.4.9b) is a consequence of the null structure. Interestingly, this is one of the two places in our proof where the null structure is used in a fundamental way.

Lemma 10.1.1. *Under the assumption of* (2.4.5) *(with $C_1\epsilon \leqslant \min(1, \epsilon_0''))$, the following estimates hold for all $|I^\sharp| \leqslant 5$ and $s_0 \leqslant s \leqslant s_1$:*

$$\|Z^{I^\sharp} F_{\hat{\imath}}\|_{L^2(\mathcal{H}_s)} \leqslant C(C_1\epsilon)^2 s^{-1+\delta}, \tag{10.1.1a}$$

$$\|[Z^I, G_i^{j\alpha\beta}(w, \partial w)\partial_\alpha\partial_\beta]w_j\| \leqslant C(C_1\epsilon)^2 s^{-1+\delta}. \tag{10.1.1b}$$

Proof. We begin with (10.1.1a). This concerns only the basic L^2 and L^∞ estimates established earlier. Recall that $Z^{I^\sharp}F_i$ is decomposed as follows:

$$Z^{I^\sharp}F_i = Z^{I^\sharp}\big(P_i^{\alpha\beta jk}\partial_\alpha w_j \partial_\beta w_k\big) + Z^{I^\sharp}\big(Q_i^{\alpha j\check{k}} v_{\check{k}}\partial_\alpha w_j\big) + Z^{I^\sharp}\big(R_i^{\check{\jmath}\check{k}} v_{\check{\jmath}} v_{\check{k}}\big)$$

We see that $Z^{I^\sharp}F_i$ is a linear combination of the following terms:

$$Z^{I^\sharp}\big(\partial_\alpha u_{\hat{\imath}}\partial_\beta u_{\hat{\jmath}}\big), \quad Z^{I^\sharp}\big(\partial_\alpha v_{\check{\imath}}\partial_\beta u_{\hat{\jmath}}\big), \quad Z^{I^\sharp}\big(\partial_\alpha v_{\check{\imath}}\partial_\beta v_{\check{\jmath}}\big),$$
$$Z^{I^\sharp}\big(v_{\check{\imath}}\partial_\alpha w_j\big), \quad Z^{I^\sharp}\big(v_{\check{\jmath}} v_{\check{k}}\big).$$

As done before, for each term and each partition of $I^\sharp = I_1^\sharp + I_2^\sharp$, we write the relevant inequalities in the following two lists (Recall our convention that, on each term, $I_1^\sharp$ acts on the first factor and $I_2^\sharp$ acts on the second, while the symbol $(a, \leqslant b)$ means $|I_1^\sharp| = a, |I_2^\sharp| \leqslant b$.):

Products	$(5, \leqslant 0)$	$(4, \leqslant 1)$	$(3, \leqslant 2)$
$\partial_\alpha u_{\hat\imath}\partial_\beta u_{\hat\jmath}$	(7.3.1a), (7.5.3a)	(7.3.1a), (7.5.3a)	(7.3.3a), (7.5.1a)
$\partial_\alpha v_{\check\imath}\partial_\beta u_{\hat\jmath}$	(7.3.1a), (7.5.3a)	(7.3.2a), (7.5.3a)	(7.3.2a), (7.5.1a)
$\partial_\alpha v_{\check\imath}\partial_\beta v_{\check\jmath}$	(7.3.1a), (7.5.2a)	(7.3.2a), (7.5.2a)	(7.3.2a), (7.5.2a)
$v_{\check\imath}\partial_\alpha w_j$	(7.3.1e), (7.5.1a)	(7.3.1e), (7.5.1a)	(7.3.1e), (7.5.1a)
$v_{\check\jmath}v_{\check k}$	(7.3.1e), (7.5.1e)	(7.3.1e), (7.5.1e)	(7.3.1e), (7.5.1e)

Products	$(2, \leqslant 3)$	$(1, \leqslant 4)$	$(0, \leqslant 5)$
$\partial_\alpha u_{\hat\imath}\partial_\beta u_{\hat\jmath}$	(7.5.1a), (7.3.3a)	(7.5.3a), (7.3.1a)	(7.5.3a), (7.3.1a)
$\partial_\alpha v_{\check\imath}\partial_\beta u_{\hat\jmath}$	(7.5.2a), (7.3.3a)	(7.5.2a), (7.3.1a)	(7.5.2a), (7.3.1a)
$\partial_\alpha v_{\check\imath}\partial_\beta v_{\check\jmath}$	(7.5.2a), (7.3.1a)	(7.5.2a), (7.3.1a)	(7.5.2a), (7.3.1a)
$v_{\check\imath}\partial_\alpha w_j$	(7.3.1e), (7.5.1a)	(7.5.1e), (7.3.1a)	(7.5.1e), (7.3.1a)
$v_{\check\jmath}v_{\check k}$	(7.3.1e), (7.5.1e)	(7.5.1e), (7.3.1e)	(7.5.1e), (7.3.1e)

From these inequalities, we conclude with (10.1.1a).

We turn our attention to (10.1.1b) and, by recalling the decomposition of $[Z^{I^\sharp}, G_i^{j\alpha\beta}(w,\partial w)\partial_\alpha\partial_\beta]w_j$, we find

$$
\begin{aligned}
&[Z^{I^\sharp}, G_i^{j\alpha\beta}(w,\partial w)\partial_\alpha\partial_\beta]w_j \\
=&[Z^{I^\sharp}, A_i^{j\alpha\beta\gamma k}\partial_\gamma w_k\partial_\alpha\partial_\beta]w_j + [Z^{I^\sharp}, B_i^{j\alpha\beta\check k}v_{\check k}\partial_\alpha\partial_\beta]w_j \\
&+ [Z^{I^\sharp}, B_i^{j\alpha\beta\hat k}u_{\hat k}\partial_\alpha\partial_\beta]u_{\hat\jmath}.
\end{aligned}
\tag{10.1.2}
$$

The first term of the right-hand side is decomposed as follows:

$$
\begin{aligned}
&[Z^{I^\sharp}, A_i^{j\alpha\beta\gamma k}\partial_\gamma w_k\partial_\alpha\partial_\beta]w_j \\
=&\sum_{\substack{I_1^\sharp+I_2^\sharp=I^\sharp\\|I_2^\sharp|\leqslant|I^\sharp|-1}} A_i^{j\alpha\beta\gamma k}Z^{I_1^\sharp}\partial_\gamma w_k Z^{I_2^\sharp}\partial_\alpha\partial_\beta w_j + A_i^{j\alpha\beta\gamma k}\partial_\gamma w_k[Z^{I^\sharp},\partial_\alpha\partial_\beta]w_j,
\end{aligned}
\tag{10.1.3}
$$

in which we have

$$
\begin{aligned}
&\sum_{\substack{I_1^\sharp+I_2^\sharp=I^\sharp\\|I_2^\sharp|\leqslant|I^\sharp|-1}} A_i^{j\alpha\beta\gamma k}Z^{I_1^\sharp}\partial_\gamma w_k Z^{I_2^\sharp}\partial_\alpha\partial_\beta w_j \\
=&\sum_{\substack{I_1^\sharp+I_2^\sharp=I^\sharp\\|I_2^\sharp|\leqslant|I^\sharp|-1}} \Big(A_i^{j\alpha\beta\gamma\hat k}Z^{I_1^\sharp}\partial_\gamma u_{\hat k}Z^{I_2^\sharp}\partial_\alpha\partial_\beta w_j + A_i^{j\alpha\beta\gamma\check k}Z^{I_1^\sharp}\partial_\gamma v_{\check k}Z^{I_2^\sharp}\partial_\alpha\partial_\beta w_j\Big).
\end{aligned}
$$

This term is a linear combination of the following terms with constant coefficients:

$$Z^{I_1^\sharp}\partial_\gamma u_{\hat{k}} Z^{I_2^\sharp}\partial_\alpha\partial_\beta u_{\hat{\jmath}}, \quad Z^{I_1^\sharp}\partial_\gamma u_{\hat{k}} Z^{I_2^\sharp}\partial_\alpha\partial_\beta v_{\check{\jmath}}$$
$$Z^{I_1^\sharp}\partial_\gamma v_{\check{k}} Z^{I_2^\sharp}\partial_\alpha\partial_\beta u_{\hat{\jmath}}, \quad Z^{I_1^\sharp}\partial_\gamma v_{\check{k}} Z^{I_2^\sharp}\partial_\alpha\partial_\beta v_{\check{\jmath}}$$

with $|I_1^\sharp| + |I_2^\sharp| \leqslant |I^\sharp|$ and $I_2^\sharp \leqslant |I^\sharp| - 1$.

The second term in the right-hand side of (10.1.3) is estimated as follows. We see that it is a linear combination of the following terms with constant coefficients:

$$\partial_\gamma u_{\hat{k}}[Z^{I^\sharp}, \partial_\alpha\partial_\beta]u_{\hat{\jmath}}, \quad \partial_\gamma u_{\hat{k}}[Z^{I^\sharp}, \partial_\alpha\partial_\beta]v_{\check{\jmath}},$$
$$\partial_\gamma v_{\check{k}}[Z^{I^\sharp}, \partial_\alpha\partial_\beta]u_{\hat{\jmath}}, \quad \partial_\gamma v_{\check{k}}[Z^{I^\sharp}, \partial_\alpha\partial_\beta]v_{\check{\jmath}}$$

By the commutator estimates (3.3.3), these terms are bounded by

$$\sum_{|I_2^\sharp|\leqslant|I^\sharp|-1} \left|\partial_\gamma u_{\hat{k}}\partial_\alpha\partial_\beta Z^{I_2^\sharp}u_{\hat{\jmath}}\right|, \quad \sum_{|I_2^\sharp|\leqslant|I^\sharp|-1} \left|\partial_\gamma u_{\hat{k}}\partial_\alpha\partial_\beta Z^{I_2^\sharp}v_{\check{\jmath}}\right|,$$
$$\sum_{|I_2^\sharp|\leqslant|I^\sharp|-1} \left|\partial_\gamma v_{\check{k}}\partial_\alpha\partial_\beta Z^{I_2^\sharp}u_{\hat{\jmath}}\right|, \quad \sum_{|I_2^\sharp|\leqslant|I^\sharp|-1} \left|\partial_\gamma v_{\check{k}}\partial_\alpha\partial_\beta Z^{I_2^\sharp}v_{\check{\jmath}}\right|.$$

We observe that $[Z^{I^\sharp}, A_i^{j\alpha\beta\gamma k}\partial_\gamma w_k\partial_\alpha\partial_\beta]w_j$ is bounded by a linear combination of the following terms with constant coefficients:

$$Z^{I_1^\sharp}\partial_\gamma u_{\hat{k}} Z^{I_2^\sharp}\partial_\alpha\partial_\beta u_{\hat{\jmath}}, \quad Z^{I_1^\sharp}\partial_\gamma u_{\hat{k}} Z^{I_2^\sharp}\partial_\alpha\partial_\beta v_{\check{\jmath}}, \quad Z^{I_1^\sharp}\partial_\gamma v_{\check{k}} Z^{I_2^\sharp}\partial_\alpha\partial_\beta u_{\hat{\jmath}},$$
$$Z^{I_1^\sharp}\partial_\gamma v_{\check{k}} Z^{I_2^\sharp}\partial_\alpha\partial_\beta v_{\check{\jmath}}, \quad Z^{I_1^\sharp}\partial_\gamma u_{\hat{k}}\partial_\alpha\partial_\beta Z^{I_2^\sharp}u_{\hat{\jmath}}, \quad Z^{I_1^\sharp}\partial_\gamma u_{\hat{k}}\partial_\alpha\partial_\beta Z^{I_2^\sharp}v_{\check{\jmath}},$$
$$Z^{I_1^\sharp}\partial_\gamma v_{\check{k}}\partial_\alpha\partial_\beta Z^{I_2^\sharp}u_{\hat{\jmath}}, \quad Z^{I_1^\sharp}\partial_\gamma v_{\check{k}}\partial_\alpha\partial_\beta Z^{I_2^\sharp}v_{\check{\jmath}}$$

with $|I_1^\sharp| + |I_2^\sharp| \leqslant |I^\sharp|$ and $|I_2^\sharp| \leqslant 4$.

We give the inequalities we use for each term and each partition of the index:

Products	$(5, \leqslant 0)$	$(4, \leqslant 1)$	$(3, \leqslant 2)$
$Z^{I_1^\sharp}\partial_\gamma u_{\hat{k}} Z^{I_2^\sharp}\partial_\alpha\partial_\beta u_{\hat{\jmath}}$	(7.3.1a), (8.4.6c)	(7.3.1b), (8.4.6b)	(7.3.3a), (8.4.6a)
$Z^{I_1^\sharp}\partial_\gamma u_{\hat{k}} Z^{I_2^\sharp}\partial_\alpha\partial_\beta v_{\check{\jmath}}$	(7.3.1a), (7.5.2a)	(7.3.1b), (7.5.2a)	(7.3.3a), (7.5.2a)
$Z^{I_1^\sharp}\partial_\gamma v_{\check{k}} Z^{I_2^\sharp}\partial_\alpha\partial_\beta u_{\hat{\jmath}}$	(7.3.1a), (8.4.6c)	(7.3.2a), (8.4.6b)	(7.3.2a), (8.4.6a)
$Z^{I_1^\sharp}\partial_\gamma v_{\check{k}} Z^{I_2^\sharp}\partial_\alpha\partial_\beta v_{\check{\jmath}}$	(7.3.1a), (7.5.1e)	(7.3.2a), (7.5.1e)	(7.3.2a), (7.5.1a)
$Z^{I_1^\sharp}\partial_\gamma u_{\hat{k}}\partial_\alpha\partial_\beta Z^{I_2^\sharp}u_{\hat{\jmath}}$	(7.3.1a), (8.4.6c)	(7.3.1b), (8.4.6b)	(7.3.3a), (8.4.6a)
$Z^{I_1^\sharp}\partial_\gamma u_{\hat{k}}\partial_\alpha\partial_\beta Z^{I_2^\sharp}v_{\check{\jmath}}$	(7.3.1a), (7.4.3a)	(7.3.1b), (7.4.3a)	(7.3.3a), (7.4.3a)
$Z^{I_1^\sharp}\partial_\gamma v_{\check{k}}\partial_\alpha\partial_\beta Z^{I_2^\sharp}u_{\hat{\jmath}}$	(7.3.1a), (8.4.6c)	(7.3.2a), (8.4.6b)	(7.3.2a), (8.4.6a)
$Z^{I_1^\sharp}\partial_\gamma v_{\check{k}}\partial_\alpha\partial_\beta Z^{I_2^\sharp}v_{\check{\jmath}}$	(7.3.1a), (7.4.2e)	(7.3.2a), (7.4.2e)	(7.3.2a), (7.4.2a)

Products	$(2, \leqslant 3)$	$(1, \leqslant 4)$
$Z^{I_1^\sharp}\partial_\gamma u_{\widehat{k}} Z^{I_2^\sharp}\partial_\alpha\partial_\beta u_{\widehat{\jmath}}$	(7.5.1b), (8.5.7b)	(7.5.3a), (8.5.7a)
$Z^{I_1^\sharp}\partial_\gamma u_{\widehat{k}} Z^{I_2^\sharp}\partial_\alpha\partial_\beta v_{\check{\jmath}}$	(7.5.1a), (7.3.2a)	(7.5.1a), (7.3.2a)
$Z^{I_1^\sharp}\partial_\gamma v_{\check{k}} Z^{I_2^\sharp}\partial_\alpha\partial_\beta u_{\widehat{\jmath}}$	(7.5.2a), (8.5.7a)	(7.5.2a), (8.5.7a)
$Z^{I_1^\sharp}\partial_\gamma v_{\check{k}} Z^{I_2^\sharp}\partial_\alpha\partial_\beta v_{\check{\jmath}}$	(7.5.2a), (7.3.1e)	(7.5.2a), (7.3.1e)
$Z^{I_1^\sharp}\partial_\gamma u_{\widehat{k}} \partial_\alpha\partial_\beta Z^{I_2^\sharp} u_{\widehat{\jmath}}$	(7.5.1b), (8.5.7b)	(7.5.3a), (8.5.7a)
$Z^{I_1^\sharp}\partial_\gamma u_{\widehat{k}} \partial_\alpha\partial_\beta Z^{I_2^\sharp} v_{\check{\jmath}}$	(7.5.1b), (7.2.2a)	(7.5.3a), (7.2.2a)
$Z^{I_1^\sharp}\partial_\gamma v_{\check{k}} \partial_\alpha\partial_\beta Z^{I_2^\sharp} u_{\widehat{\jmath}}$	(7.5.2a), (8.5.7a)	(7.5.2a), (8.5.7a)
$Z^{I_1^\sharp}\partial_\gamma v_{\check{k}} \partial_\alpha\partial_\beta Z^{I_2^\sharp} v_{\check{\jmath}}$	(7.5.2a), (7.2.1e)	(7.5.2a), (7.2.1e)

The remaining terms in the right-hand side of (10.1.2) are estimated similarly. These two terms can be bounded by the combination of the following terms with constant coefficients:

$$Z^{I_1^\sharp} v_{\check{k}} Z^{I_2^\sharp}\partial_\alpha\partial_\beta w_j, \quad Z^{I_1^\sharp} u_{\widehat{k}} Z^{I_2^\sharp}\partial_\alpha\partial_\beta u_{\widehat{\jmath}},$$
$$Z^{I_1^\sharp} v_{\check{k}} \partial_\alpha\partial_\beta Z^{I_2^\sharp} w_j, \quad Z^{I_1^\sharp} u_{\widehat{k}} \partial_\alpha\partial_\beta Z^{I_2^\sharp} u_{\widehat{\jmath}}$$

with $|I_1^\sharp| + |I_2^\sharp| \leqslant |I^\sharp|$ and $|I_2^\sharp| \leqslant |I^\sharp| - 1$. We will write in details the estimate for the most critical term $Z^{I_1^\sharp} u_{\widehat{k}} Z^{I_2^\sharp}\partial_\alpha\partial_\beta u_{\widehat{\jmath}}$:

$$\begin{aligned}
&\sum_{\substack{|I_1^\sharp|+|I_2^\sharp|\leqslant|I^\sharp|\\ |I_2^\sharp|\leqslant|I^\sharp|-1}} \big|Z^{I_1^\sharp} u_{\widehat{k}} Z^{I_2^\sharp}\partial_\alpha\partial_\beta u_{\widehat{\jmath}}\big| \\
&\leqslant \sum_{|I_1^\sharp|\leqslant 5} \big|Z^{I_1^\sharp} u_{\widehat{k}} \partial_\alpha\partial_\beta u_{\widehat{\jmath}}\big| + \sum_{|I_1^\sharp|=4, |I_2^\sharp|\leqslant 1} \big|Z^{I_1^\sharp} u_{\widehat{k}} Z^{I_2^\sharp}\partial_\alpha\partial_\beta u_{\widehat{\jmath}}\big| \\
&\quad + \sum_{|I_1^\sharp|=3, |I_2^\sharp|\leqslant 2} \big|Z^{I_1^\sharp} u_{\widehat{k}} Z^{I_2^\sharp}\partial_\alpha\partial_\beta u_{\widehat{\jmath}}\big| \\
&\quad + \sum_{|I_1^\sharp|=2, |I_2^\sharp|\leqslant 3} \big|Z^{I_1^\sharp} u_{\widehat{k}} Z^{I_2^\sharp}\partial_\alpha\partial_\beta u_{\widehat{\jmath}}\big| + \sum_{|I_1^\sharp|=1, |I_2^\sharp|\leqslant 4} \big|Z^{I_1^\sharp} u_{\widehat{k}} Z^{I_2^\sharp}\partial_\alpha\partial_\beta u_{\widehat{\jmath}}\big| \\
&=: T_5 + T_4 + T_3 + T_2 + T_1.
\end{aligned}$$

The term T_5 is estimated by applying (7.3.6a) on the first factor and (8.4.6c) on the second factor:

$$\begin{aligned}
\|T_5\|_{L^2(\mathcal{H}_s)} &= \sum_{|I_1^\sharp|\leqslant 5} \big\|s^{-1} Z^{I_1^\sharp} u_{\widehat{k}}\, s\partial_\alpha\partial_\beta u_{\widehat{\jmath}}\big\|_{L^2(\mathcal{H}_s)} \\
&\leqslant \big\|s^{-1} Z^{I_1^\sharp} u_{\widehat{k}}\big\|_{L^2(\mathcal{H}_s)} \big\|s\partial_\alpha\partial_\beta u_{\widehat{\jmath}}\big\|_{L^\infty(\mathcal{H}_s)} \\
&\leqslant C(C_1\epsilon)^2 s^\delta \big\|t^{1/2} s^{-2}\big\|_{L^\infty(\mathcal{H}_s)} \leqslant C(C_1\epsilon)^2 s^{-1+\delta},
\end{aligned}$$

where we observe that $s \leqslant t \leqslant Cs^2$ in the half-cone $\mathcal{K}$.

The term T_4 is bounded by applying (7.3.6b) and (8.4.6b):

$$\begin{aligned}\|T_4\|_{L^2(\mathcal{H}_s)} \leqslant & \sum_{|I_1^\sharp|=4,|I_2^\sharp|\leqslant 1} \big\|s^{-1} Z^{I_1^\sharp} u_{\widehat{k}}\, s Z^{I_2^\sharp}\partial_\alpha\partial_\beta u_{\widehat{\jmath}}\big\|_{L^2(\mathcal{H}_s)} \\ \leqslant & C(C_1\epsilon)^2 s^{\delta/2}\big\|st^{1/2}s^{-3+\delta/2}\big\|_{L^\infty(\mathcal{H}_s)} \\ \leqslant & C(C_1\epsilon)s^{-1+\delta}.\end{aligned}$$

The term T_3 is bounded by applying (7.5.5a) and (8.5.7c):

$$\begin{aligned}\|T_3\|_{L^2(\mathcal{H}_s)} \leqslant & \sum_{|I_1^\sharp|=3,|I_2^\sharp|\leqslant 2} \big\|s^{-1}t^{3/2} Z^{I_1^\sharp} u_{\widehat{k}}\, st^{-3/2} Z^{I_2^\sharp}\partial_\alpha\partial_\beta u_{\widehat{\jmath}}\big\|_{L^2(\mathcal{H}_s)} \\ \leqslant & CC_1\epsilon s^{\delta}\big\|s^3t^{-2}Z^{I_2^\sharp}\partial_\alpha\partial_\beta u_{\widehat{\jmath}}\big\|_{L^2(\mathcal{H}_s)}\big\|s^{-2}t^{1/2}\big\|_{L^\infty(\mathcal{H}_s)} \\ \leqslant & C(C_1\epsilon)^2 s^{-1+\delta}.\end{aligned}$$

The term T_2 is estimated by applying (7.5.5b), (8.5.7b):

$$\begin{aligned}\|T_2\|_{L^2(\mathcal{H}_s)} \leqslant & \sum_{|I_1^\sharp|\leqslant 2,|I_2^\sharp|\leqslant 3} \big\|s^{-3}t^2 Z^{I_1^\sharp} u_{\widehat{k}}\, s^3t^{-2} Z^{I_2^\sharp}\partial_\alpha\partial_\beta u_{\widehat{\jmath}}\big\|_{L^2(\mathcal{H}_s)} \\ \leqslant & \sum_{|I_1^\sharp|\leqslant 2,|I_2^\sharp|\leqslant 3} \big\|s^{-3}t^2 Z^{I_1^\sharp} u_{\widehat{k}}\big\|_{L^\infty(\mathcal{H}_s)}\big\|s^3t^{-2} Z^{I_2^\sharp}\partial_\alpha\partial_\beta u_{\widehat{\jmath}}\big\|_{L^2(\mathcal{H}_s)} \\ \leqslant & CC_1\epsilon\big\|s^{-3}t^2t^{-3/2}s^{1+\delta/2}\big\|_{L^\infty(\mathcal{H}_s)}CC_1\epsilon s^{\delta/2} \\ = & C(C_1\epsilon)^2 s^{-1+\delta}.\end{aligned}$$

The term T_1 is bounded by applying (7.5.5c) and (8.5.7a):

$$\begin{aligned}\|T_1\|_{L^2(\mathcal{H}_s)} \leqslant & \sum_{|I_1^\sharp|\leqslant 1,|I_2^\sharp|\leqslant 4} \big\|s^{-3}t^2 Z^{I_1^\sharp} u_{\widehat{k}}\, s^3t^{-2} Z^{I_2^\sharp}\partial_\alpha\partial_\beta u_{\widehat{\jmath}}\big\|_{L^2(\mathcal{H}_s)} \\ \leqslant & \sum_{|I_1^\sharp|\leqslant 1,|I_2^\sharp|\leqslant 4} \big\|s^{-3}t^2 Z^{I_1^\sharp} u_{\widehat{k}}\big\|_{L^\infty(\mathcal{H}_s)}\big\|s^3t^{-2} Z^{I_2^\sharp}\partial_\alpha\partial_\beta u_{\widehat{\jmath}}\big\|_{L^2(\mathcal{H}_s)} \\ \leqslant & CC_1\epsilon\|s^{-3}t^2t^{-3/2}s\|_{L^\infty(\mathcal{H}_s)}\cdot CC_1\epsilon s^{\delta} \leqslant C(C_1\epsilon)^2 s^{-1+\delta}.\end{aligned}$$

For the remaining terms, we list out the inequalities to be used on each term and each partition of the index as follows:

Products	$(5, \leqslant 0)$	$(4, \leqslant 1)$	$(3, \leqslant 2)$
$Z^{I_1^\sharp}v_{\widetilde{k}}Z^{I_2^\sharp}\partial_\alpha\partial_\beta w_j$	(7.3.1e), (7.5.1a)	(7.3.1e), (7.5.1a)	(7.3.1e), (7.5.1a)
$Z^{I_1^\sharp}u_{\widehat{k}}Z^{I_2^\sharp}\partial_\alpha\partial_\beta u_{\widehat{\jmath}}$	(7.3.6a), (8.4.6c)	(7.3.6b), (8.4.6b)	(7.5.5a), (8.5.7c)
$Z^{I_1^\sharp}v_{\widetilde{k}}\partial_\alpha\partial_\beta Z^{I_2^\sharp}w_j$	(7.3.1e), (7.4.2a)	(7.3.1e), (7.4.2a)	(7.3.1e), (7.4.2a)
$Z^{I_1^\sharp}u_{\widehat{k}}\partial_\alpha\partial_\beta Z^{I_2^\sharp}u_{\widehat{\jmath}}$	(7.3.6a), (8.4.6c)	(7.3.6b), (8.4.6b)	(7.5.5a), (8.5.7c)

Products	$(2, \leqslant 3)$	$(1, \leqslant 4)$
$Z^{I_1^\sharp} v_{\check{k}} Z^{I_2^\sharp} \partial_\alpha \partial_\beta w_j$	(7.5.1e), (7.3.1a)	(7.5.1e), (7.3.1a)
$Z^{I_1^\sharp} u_{\hat{k}} Z^{I_2^\sharp} \partial_\alpha \partial_\beta u_{\hat{\jmath}}$	(7.5.5b), (8.5.7b)	(7.5.5c), (8.5.7a)
$Z^{I_1^\sharp} v_{\check{k}} \partial_\alpha \partial_\beta Z^{I_2^\sharp} w_j$	(7.5.1e), (7.2.1a)	(7.5.1e), (7.2.1a)
$Z^{I_1^\sharp} u_{\hat{k}} \partial_\alpha \partial_\beta Z^{I_2^\sharp} u_{\hat{\jmath}}$	(7.5.5b), (8.5.7b)	(7.5.5c), (8.5.7a)

□

We estimate the source with fourth-order derivatives.

Lemma 10.1.2. *By relying on the assumption* (2.4.5), *the following estimates hold for all* $|I^\dagger| \leqslant 4$ *and* $s_0 \leqslant s \leqslant s_1$*:*

$$\|Z^{I^\dagger} F_{\hat{\imath}}\|_{L^2(\mathcal{H}_s)} \leqslant C(C_1\epsilon)^2 s^{-1+\delta/2}, \tag{10.1.4a}$$

$$\|[Z^{I^\dagger}, G_i^{j\alpha\beta}(w, \partial w)\partial_\alpha\partial_\beta] w_j\|_{L^2(\mathcal{H}_s)} \leqslant C(C_1\epsilon)^2 s^{-1+\delta/2}. \tag{10.1.4b}$$

Proof. The proof is essentially the same as the one of Lemma 10.1.1. We will not write the proof in detail, but we list out the inequalities we use for each term and each partition of the index.

First we list out the estimate on $Z^{I^\dagger} F_i$ in Table 15. The terms $[Z^{I^\dagger}, B_i^{j\alpha\beta k} w_k \partial_\alpha \partial_\beta] w_j$ are estimated by the inequalities listed in Table 16. □

10.2 L^2 estimates on third-order terms

We now derive the L^2 estimates for the source terms (2.4.9b). This is the second place where the null structure is taken into consideration after Lemma 9.3.4.

Lemma 10.2.1. *By relying on the assumption of* (2.4.5) *with* $C_1\epsilon < \min(1, \epsilon_0'')$, *the following estimates hold for all* $|I| \leqslant 3$*:*

$$\|Z^I F_{\hat{\imath}}\|_{L^2(\mathcal{H}_s)} \leqslant C(C_1\epsilon)^2 s^{-3/2+2\delta}, \tag{10.2.1a}$$

$$\|Z^I G_{\hat{\imath}}^{\check{\jmath}\alpha\beta}(w, \partial w)\partial_\alpha\partial_\beta v_{\check{\jmath}}\|_{L^2(\mathcal{H}_s)} \leqslant C(C_1\epsilon)^2 s^{-3/2+2\delta}, \tag{10.2.1b}$$

$$\|[Z^I, G_{\hat{\imath}}^{\hat{\jmath}\alpha\beta}(w, \partial w)\partial_\alpha\partial_\beta] u_{\hat{\jmath}}\|_{L^2(\mathcal{H}_s)} \leqslant C(C_1\epsilon)^2 s^{-3/2+2\delta}. \tag{10.2.1c}$$

Table 15

Products	$(4, \leqslant 0)$	$(3, \leqslant 1)$	$(2, \leqslant 2)$	$(1, \leqslant 3)$	$(0, \leqslant 4)$
$\partial_\alpha u_{\hat{\imath}} \partial_\beta u_{\hat{\jmath}}$	(7.3.1b), (7.5.3a)	(7.3.3a), (7.5.3a)	(7.3.3a), (7.5.1b)	(7.5.3a), (7.3.3a)	(7.5.3a), (7.3.1b)
$\partial_\alpha v_{\check{\imath}} \partial_\beta w_j$	(7.3.2a), (7.5.1a)	(7.3.2a), (7.5.1a)	(7.3.2a), (7.5.1a)	(7.5.2a), (7.3.1a)	(7.5.2a), (7.3.1a)
$v_{\check{\imath}} \partial_\alpha w_j$	(7.3.1e), (7.5.1a)	(7.3.1e), (7.5.1a)	(7.3.1e), (7.5.1a)	(7.5.1e), (7.3.1a)	(7.5.1e), (7.3.1a)
$v_{\check{\jmath}} v_{\check{k}}$	(7.3.1e), (7.5.1e)	(7.3.1e), (7.5.1e)	(7.3.1e), (7.5.1e)	(7.5.1e), (7.3.1e)	(7.5.1e), (7.3.1e)

Table 16

Products	$(4, \leqslant 0)$	$(3, \leqslant 1)$	$(2, \leqslant 2)$	$(1, \leqslant 3)$
$Z^{I_1^\dagger} \partial_\gamma u_{\hat{k}} Z^{I_2^\dagger} \partial_\alpha \partial_\beta u_{\hat{\jmath}}$	(7.3.1b), (8.4.6c)	(7.3.3a), (8.4.6b)	(7.5.1b), (8.5.7c)	(7.5.3a), (8.5.7b)
$Z^{I_1^\dagger} \partial_\gamma u_{\hat{k}} \partial_\alpha \partial_\beta Z^{I_2^\dagger} u_{\hat{\jmath}}$	(7.3.1b), (8.4.6c)	(7.3.3a), (8.4.6b)	(7.5.1b), (8.5.7c)	(7.5.3a), (8.5.7b)
$Z^{I_1^\dagger} \partial_\gamma u_{\hat{k}} Z^{I_2^\dagger} \partial_\alpha \partial_\beta v_{\check{\jmath}}$	(7.3.1b), (7.5.2a)	(7.3.3a), (7.5.2a)	(7.5.1b), (7.3.2b)	(7.5.3a), (7.3.2a)
$Z^{I_1^\dagger} \partial_\gamma u_{\hat{k}} \partial_\alpha \partial_\beta Z^{I_2^\dagger} v_{\check{\jmath}}$	(7.3.1b), (7.4.3a)	(7.3.3a), (7.4.3a)	(7.5.1b), (7.2.2b)	(7.5.3a), (7.2.2a)
$Z^{I_1^\dagger} \partial_\gamma v_{\check{k}} Z^{I_2^\dagger} \partial_\alpha \partial_\beta w_j$	(7.3.2a), (7.5.1a)	(7.3.2a), (7.5.1a)	(7.5.2a), (7.3.1a)	(7.5.2a), (7.3.1a)
$Z^{I_1^\dagger} \partial_\gamma v_{\check{k}} \partial_\alpha \partial_\beta Z^{I_2^\dagger} w_j$	(7.3.2a), (7.5.1a)	(7.3.2a), (7.5.1a)	(7.5.2a), (7.2.1a)	(7.5.2a), (7.2.1a)
$Z^{I_1^\dagger} v_{\check{k}} Z^{I_2^\dagger} \partial_\alpha \partial_\beta w_j$	(7.3.1e), (7.5.1a)	(7.3.1e), (7.5.1a)	(7.5.1e), (7.3.1a)	(7.5.1e), (7.3.1a)
$Z^{I_1^\dagger} v_{\check{k}} \partial_\alpha \partial_\beta Z^{I_2^\dagger} w_j$	(7.3.1e), (7.5.1e)	(7.3.1e), (7.5.1e)	(7.5.1e), (7.2.1e)	(7.5.1e), (7.2.1e)
$Z^{I_1^\dagger} u_{\hat{k}} Z^{I_2^\dagger} \partial_\alpha \partial_\beta u_{\hat{\jmath}}$	(7.3.6b), (8.4.6c)	(7.3.6c), (8.4.6b)	(7.5.5b), (8.5.7c)	(7.5.5c), (8.5.7b)
$Z^{I_1^\dagger} u_{\hat{k}} \partial_\alpha \partial_\beta Z^{I_2^\dagger} u_{\hat{\jmath}}$	(7.3.6b), (8.4.6c)	(7.3.6c), (8.4.6b)	(7.5.5b), (8.5.7c)	(7.5.5c), (8.5.7b)

Proof. We first prove (10.2.1a) and recall the structure of $Z^I F_{\widehat{\imath}}$:

$$
\begin{aligned}
Z^I F_{\widehat{\imath}} = & Z^I\big(P_{\widehat{\imath}}^{\alpha\beta\widehat{\jmath}\widehat{k}}\partial_\alpha u_{\widehat{\jmath}}\partial_\beta u_{\widehat{k}}\big)\\
& + P_{\widehat{\imath}}^{\alpha\beta\check{\jmath}\widehat{k}} Z^I\big(\partial_\alpha v_{\check{\jmath}}\partial_\beta u_{\widehat{k}}\big) + P_{\widehat{\imath}}^{\alpha\beta\widehat{\jmath}\check{k}} Z^I\big(\partial_\alpha u_{\widehat{\jmath}}\partial_\beta v_{\check{k}}\big) + P_{\widehat{\imath}}^{\alpha\beta\check{\jmath}\check{k}} Z^I\big(\partial_\alpha v_{\check{\jmath}}\partial_\beta v_{\check{k}}\big)\\
& + Q_{\widehat{\imath}}^{\alpha j\check{k}} Z^I\big(v_{\check{k}}\partial_\alpha w_j\big) + R_{\widehat{\imath}}^{\check{\jmath}\check{k}} Z^I\big(v_{\check{\jmath}}v_{\check{k}}\big).
\end{aligned}
$$

The first term in the right-hand side is a null term and we can apply directly (9.1.1a):

$$
\big\|P_{\widehat{\imath}}^{\alpha\beta\widehat{\jmath}\widehat{k}}\partial_\alpha u_{\widehat{\jmath}}\partial_\beta u_{\widehat{k}}\big\|_{L^2(\mathcal{H}_s)} \leqslant C(C_1\epsilon)^2 s^{-3/2}.
$$

The remaining terms are linear combinations of the following terms with constant coefficients:

$$
Z^I\big(\partial_\alpha v_{\check{\jmath}}\partial_\beta w_k\big), \quad Z^I\big(v_{\check{k}}\partial_\alpha w_j\big), \quad Z^I\big(v_{\check{\jmath}}v_{\check{k}}\big).
$$

We will not give in details the estimates of each term, but we list out the inequalities we use for each term and each partition:

Products	$(3,\leqslant 0)$	$(2,\leqslant 1)$	$(1,\leqslant 2)$	$(0,\leqslant 3)$
$\partial_\alpha v_{\check{\jmath}}\partial_\beta w_k$	(7.3.2a), (7.5.1a)	(7.3.2a), (7.5.1a)	(7.5.2a), (7.3.1a)	(7.5.2a), (7.3.1a)
$v_{\check{k}}\partial_\alpha w_j$	(7.3.1e), (7.5.1a)	(7.3.1e), (7.5.1a)	(7.5.1e), (7.3.1a)	(7.5.1e), (7.3.1a)
$v_{\check{\jmath}}v_{\check{k}}$	(7.3.1e), (7.5.1e)	(7.3.1e), (7.5.1e)	(7.5.1e), (7.3.1e)	(7.5.1e), (7.3.1e)

To establish the second inequality (10.2.1b), we decompose it as follows:

$$
Z^I\big(G_{\widehat{\imath}}^{\check{\jmath}\alpha\beta}(w,\partial w)\partial_\alpha\partial_\beta v_{\check{\jmath}}\big) = A_{\widehat{\imath}}^{\check{\jmath}\alpha\beta\gamma k} Z^I\big(\partial_\gamma w_k\partial_\alpha\partial_\beta v_{\check{\jmath}}\big) + B_{\widehat{\imath}}^{\check{\jmath}\alpha\beta\check{k}} Z^I\big(v_{\check{k}}\partial_\alpha\partial_\beta v_{\check{\jmath}}\big).
$$

So, it is to be bounded by a linear combination of the following terms with constant coefficients:

$$
Z^I\big(\partial_\gamma w_k\partial_\alpha\partial_\beta v_{\check{\jmath}}\big), \quad Z^I\big(v_{\check{k}}\partial_\alpha\partial_\beta v_{\check{\jmath}}\big).
$$

As before, we list out the inequalities we use for each term and each partition of the index:

Products	$(3,\leqslant 0)$	$(2,\leqslant 1)$	$(1,\leqslant 2)$	$(0,\leqslant 3)$
$\partial_\gamma w_k\partial_\alpha\partial_\beta v_{\check{\jmath}}$	(7.3.1a), (7.5.2a)	(7.3.1a), (7.5.2a)	(7.5.1a), (7.3.2a)	(7.5.1a), (7.3.2a)
$v_{\check{k}}\partial_\alpha\partial_\beta v_{\check{\jmath}}$	(7.3.1e), (7.5.2a)	(7.3.1e), (7.5.2a)	(7.5.1e), (7.3.2a)	(7.5.1e), (7.3.2a)

The proof of the inequality (10.2.1c) is also related to the null structure. Recall the following decomposition:

$$
\begin{aligned}
& [Z^I, G_{\widehat{\imath}}^{\widehat{\jmath}\alpha\beta}(w,\partial w)\partial_\alpha\partial_\beta]u_{\widehat{\jmath}}\\
=& [Z^I, A_{\widehat{\imath}}^{\widehat{\jmath}\alpha\beta\gamma\widehat{k}}\partial_\gamma u_{\widehat{k}}\partial_\alpha\partial_\beta]u_{\widehat{\jmath}} + [Z^I, B_{\widehat{\imath}}^{\widehat{\jmath}\alpha\beta\widehat{k}} u_{\widehat{k}}\partial_\alpha\partial_\beta]u_{\widehat{\jmath}}\\
& + [Z^I, A_{\widehat{\imath}}^{\widehat{\jmath}\alpha\beta\gamma\check{k}}\partial_\gamma v_{\check{k}}\partial_\alpha\partial_\beta]u_{\widehat{\jmath}} + [Z^I, B_{\widehat{\imath}}^{jh\alpha\beta\check{k}} v_{\check{k}}\partial_\alpha\partial_\beta]u_{\widehat{\jmath}}
\end{aligned}
$$

Observing the null structure of the first two terms and applying (9.1.1b) and (9.2.1a), we obtain

$$\begin{aligned}&\big\|[Z^I, A_{\hat{\imath}}^{\hat{\jmath}\alpha\beta\gamma\hat{k}}\partial_\gamma u_{\hat{k}}\partial_\alpha\partial_\beta]u_{\hat{\jmath}}\big\|_{L^2(\mathcal{H}_s)} + \big\|[Z^I, B_{\hat{\imath}}^{\hat{\jmath}\alpha\beta\hat{k}}u_{\hat{k}}\partial_\alpha\partial_\beta]u_{\hat{\jmath}}\big\|_{L^2(\mathcal{H}_s)}\\ &\leqslant C(C_1\epsilon)^2 s^{-3/2+\delta}.\end{aligned}$$

The remaining two terms are linear combinations of the following term:

$$\begin{aligned}&Z^{I_1}\partial_\gamma v_{\check{k}} Z^{I_2}\partial_\alpha\partial_\beta u_{\hat{\jmath}}, \quad Z^{I_1} v_{\check{k}} Z^{I_2}\partial_\alpha\partial_\beta u_{\hat{\jmath}},\\ &Z^{I_1}\partial_\gamma v_{\check{k}}\partial_\alpha\partial_\beta Z^{I_2} u_{\hat{\jmath}}, \quad Z^{I_1} v_{\check{k}}\partial_\alpha\partial_\beta Z^{I_2} u_{\hat{\jmath}}\end{aligned}$$

with $|I_1| + |I_2| \leqslant |I|$ and $|I_2| \leqslant 2$.

We omit here the details and list out the inequalities applied to each term and each partition of the index:

Products	$(3, \leqslant 0)$	$(2, \leqslant 1)$	$(1, \leqslant 2)$
$Z^{I_1}\partial_\gamma v_{\check{k}} Z^{I_2}\partial_\alpha\partial_\beta u_{\hat{\jmath}}$	(7.3.2a), (7.5.1a)	(7.3.2a), (7.5.1a)	(7.5.2a), (7.3.1a)
$Z^{I_1} v_{\check{k}} Z^{I_2}\partial_\alpha\partial_\beta u_{\hat{\jmath}}$	(7.3.1e), (7.5.1a)	(7.3.1e), (7.5.1a)	(7.5.1e), (7.3.1a)
$Z^{I_1}\partial_\gamma v_{\check{k}}\partial_\alpha\partial_\beta Z^{I_2} u_{\hat{\jmath}}$	(7.3.2a), (7.4.2a)	(7.3.2a), (7.4.2a)	(7.5.2a), (7.2.1a)
$Z^{I_1} v_{\check{k}}\partial_\alpha\partial_\beta Z^{I_2} u_{\hat{\jmath}}$	(7.3.1e), (7.4.2a)	(7.3.1e), (7.4.2a)	(7.5.1e), (7.2.1a)

□

Chapter 11

The local well-posedness theory

11.1 Construction of the initial data

In this chapter, we establish the local-in-time existence theory for general systems of the form

$$\begin{aligned}
&\Box w_i + G_i^{j\alpha\beta}(w, \partial w)\partial_\alpha \partial_\beta w_j + c_i^2 w_i = F_i(w, \partial w), \\
&w_i(B+1, x) = w_{i0}, \quad \partial_t w_i(B+1, x) = w_{i1},
\end{aligned} \tag{11.1.1}$$

where

$$\begin{aligned}
G_i^{j\alpha\beta}(w, \partial w) &= A_i^{j\alpha\beta\gamma k}\partial_\gamma w_k + B^{j\alpha\beta k} w_k, \\
F_i(w, \partial w) &= P_i^{\alpha\beta jk}\partial_\alpha w_j \partial_\beta w_k + Q_i^{\alpha jk}\partial_\alpha w_j w_k + R_i^{jk} w_j w_k,
\end{aligned} \tag{11.1.2}$$

with constants $A_i^{\alpha\beta\gamma j}, B^{j\alpha\beta k}, P_i^{\alpha\beta jk}, Q_i^{\alpha jk}, R_i^{jk}$. To guarantee the hyperbolicity property, the following symmetry conditions are assumed:

$$G_i^{j\alpha\beta} = G_j^{i\alpha\beta}, \qquad G_i^{j\alpha\beta} = G_i^{j\beta\alpha}. \tag{11.1.3}$$

Initial data by $(w_{i0}, w_{i1}) \in \mathbf{H}^{l+1}(\mathbb{R}^3) \times \mathbf{H}^l(\mathbb{R}^3)$ of sufficiently high regularity are prescribed on the hyperplane $\{t = B+1\}$ and we assume the smallness condition:

$$\|w_{i0}\|_{\mathbf{H}^{l+1}(\mathbb{R}^3)} + \|w_{i1}\|_{\mathbf{H}^l(\mathbb{R}^3)} \leqslant \epsilon'. \tag{11.1.4}$$

We also assume that w_{i0} and w_{i1} are supported in the ball $\{|x| \leqslant B\}$. The following theorem was essentially already established in Sogge (2008).

Theorem 11.1.1. *For any integer $l \geqslant 5$, there exists a time interval $[B+1, T(\epsilon') + B + 1]$ on which the Cauchy problem (11.1.1) admits a (unique) solution $w_i = w_i(t,x)$ and one has*

$$w_i(t,x) \in C\big([B+1, T(\epsilon) + B + 1], \boldsymbol{H}^{l+1}(\mathbb{R}^3)\big) \tag{11.1.5}$$

and there exists a constant $A > 0$ such that

$$\sum_i \sum_{\substack{\alpha \\ |I| \leqslant 5}} \|\partial_\alpha Z^I w_i\|_{L^\infty([B+1,T(\epsilon)],L^2(\mathbb{R}^3))} \leqslant A\epsilon'. \tag{11.1.6}$$

Moreover, the time of existence approaches infinity when the size of the data approaches zero, that is,

$$\lim_{\epsilon' \to 0^+} T(\epsilon') = +\infty. \tag{11.1.7}$$

If T^* denotes the supremum of all such times $T(\epsilon')$ (for fixed initial data), then either $T^* = +\infty$ or else

$$\sup_{\substack{t \in [0,T^*] \\ x \in \mathbb{R}}} \sum_i \sum_{|\beta| \leqslant 5} |\partial^\beta w_i(t,x)| = +\infty. \tag{11.1.8}$$

In addition, for $C_0 > 1$ sufficiently large and some $\epsilon_0' > 0$ (depending only on the structure of the system), the uniform bound

$$\sum_i E_G(B+1, w_i) \leqslant C_0 \epsilon' \tag{11.1.9}$$

holds for all initial data satisfying $\epsilon' \leqslant \epsilon_0'$, where $E_G(\cdot, w_i) := E_{G,0}(\cdot, w_i)$ is the hyperboloidal energy in (2.3.4).

Proof. We only sketh the proof. The local existence theory and the blow-up criteria are discussed in Sogge (2008) (cf. by Theorem 4.1 in Section 1.4 therein). The property of $T(\epsilon')$ is deduced from our proof of global existence in this monograph. Therefore, we can focus here on estimating the energy $E_G(B+1, Z^I w_i)$.

We consider the region $\mathcal{K}_{B+1} := \{(t,x) : t \geqslant B+1, t^2 - |x|^2 \leqslant (B+1)^2\} \cap \mathcal{K}$ where $\mathcal{K} = \{|x| < t-1, t \geqslant 0\}$. We observe that in $\mathcal{K}_{B+1}$, $t \leqslant \frac{(B+1)^2+1}{2}$. We can fix some ϵ_0' sufficiently small, so that $T(\epsilon') \geqslant \frac{(B+1)^2+1}{2}$. The local solution is well defined in the region $\mathcal{K}_{B+1}$.

To estimate $E_G(B+1, Z^I w_i)$, we choose $\partial_t w_i$ as a multiplier and, with the notation

$$\begin{aligned} &E_G^*(B+1, w_i) \\ &:= \int_{\mathbb{R}^3} \Big(\sum_\alpha |\partial_\alpha w_i|^2 + 2G_i^{j\alpha\beta} \partial_t w_i \partial_\beta w_j - G_i^{j\alpha\beta} \partial_\alpha w_i \partial_\beta w_j\Big)(B+1, \cdot)\, dx \end{aligned} \tag{11.1.10}$$

in order to denote the standard energy defined on the flat hypersurface $t = B+1$, we obtain the energy estimate

$$
\begin{aligned}
&\sum_i E_G(B+1, Z^I w_i) - \sum_i E_G^*(B+1, Z^I w_i) \\
&= \sum_i \int_{\mathcal{K}_{B+1}} \big(Z^I F_i(w, \partial w)\, \partial_t Z^I w_i - [Z^I, G_i^{j\alpha\beta} \partial_{\alpha\beta}] Z^I w_j\, \partial_t Z^I w_i\big) dxdt \\
&\quad + \sum_i \int_{\mathcal{K}_{B+1}} \Big(\partial_\alpha G_i^{j\alpha\beta} \partial_t Z^I w_i \partial_\beta Z^I w_j - \frac{1}{2} \partial_t G_i^{j\alpha\beta} \partial_\alpha Z^I w_i \partial_\beta Z^I w_j\Big) dxdt \\
&\leqslant CK \sum_{\substack{i,j,k \\ \alpha,\beta}} \int_{B+1}^{\frac{(B+1)^2+1}{2}} |\partial_\alpha Z^I w_i|\, |\partial_\beta Z^J w_j|\, |\partial_\gamma Z^{J'} w_k| dxdt \leqslant CA\epsilon'.
\end{aligned}
$$

Here, C is a constant depending only on the structure of the system. We have

$$E_G(B+1, Z^I w_i) \leqslant CA\epsilon' + C\epsilon'.$$

Thus, for $C_0 > 1$ sufficiently large and $\epsilon' > 0$, we can find some ϵ_0' sufficiently small such that

$$CA\epsilon' + C\epsilon' \leqslant C_0\epsilon'.$$

□

11.2 Local well-posedness theory in the hyperboloidal foliation

For convenience, we introduce the following notation in the cone $\mathcal{K}$

$$
\begin{aligned}
\bar{x}^0 &:= s = \sqrt{t^2 - |x|^2}, \\
\bar{x}^a &:= x^a.
\end{aligned}
$$

The natural frame associated with these variables is

$$
\begin{aligned}
\bar{\partial}_0 &= \partial_s = \frac{s}{t}\partial_t, \\
\bar{\partial}_a &= \underline{\partial}_a = \frac{x^a}{t}\partial_t + \partial_a.
\end{aligned}
$$

By an easy calculation, we express the wave operator in this frame:

$$\Box u = \bar{\partial}_0 \bar{\partial}_0 u + \frac{2\bar{x}^a}{s} \bar{\partial}_0 \bar{\partial}_a u - \sum_a \bar{\partial}_a \bar{\partial}_a u + \frac{3}{s} \underline{\partial}_a u. \tag{11.2.1}$$

The symmetric second-order quasi-linear operator $G_i^{j\alpha\beta}(w,\partial w)\partial_\alpha\partial_\beta w_j$ can also be expressed in this frame:

$$\begin{aligned} & G_i^{j\alpha\beta}(w,\partial w)\partial_\alpha\partial_\beta w_j \\ & = \bar{G}_i^{j\alpha\beta}(w,\bar{\partial} w)\partial_\alpha\partial_\beta w_j + G_i^{j\alpha\beta}(w,\partial w)\partial_\alpha\big(\Psi_\beta^{\beta'}\big)\bar{\partial}_{\beta'} w_j \end{aligned} \tag{11.2.2}$$

with

$$\bar{G}_i^{j\alpha\beta} = G_i^{j\alpha'\beta'}\Psi_{\alpha'}^{\alpha}\Psi_{\beta'}^{\beta},$$

which is again symmetric with respect to i,j.

Now, we can transform the system (11.1.1):

$$\begin{aligned} \bar{\partial}_0\bar{\partial}_0 w_i + \bar{G}_i^{j00}\bar{\partial}_0\bar{\partial}_0 w_j + \frac{\bar{x}^a}{s}\bar{\partial}_0\bar{\partial}_a w_i + \bar{G}_i^{j0a}(w,\partial w)\bar{\partial}_0\bar{\partial}_a w_j & \\ - \sum_a \bar{\partial}_a\bar{\partial}_a w_i + \bar{G}_i^{jab}\bar{\partial}_a\bar{\partial}_b w_j = \tilde{F}_i(w,\partial w) & \end{aligned} \tag{11.2.3}$$

with initial data posed on $s = B+1$:

$$w_i(B+1,\cdot) = w_{i0}', \quad \partial_s w_i(B+1,\cdot) = w_{i1}'.$$

We observe that this system is again symmetric and that the perturbation terms $\bar{G}$ are small, so this system is again hyperbolic with respect to the variables $(s,\bar{x}^a)$. By the standard theory of second-order symmetric hyperbolic system, the following theorem holds. (See, for example, Taylor (2011), Proposition 3.1 in Chap. 16, Section 3.)

Theorem 11.2.1. *Let $l \geqslant 5$ be an integer and $B > 0$. Then, the following property holds for a sufficiently small $\epsilon > 0$. Let $(w_{i0}, w_{i1}) \in \boldsymbol{H}^{l+1} \times \boldsymbol{H}^{l}$ be data supported in the ball centered at the origin and with radius $\frac{(B+1)^2-1}{2}$. Then, the Cauchy problem*

$$\begin{aligned} & \Box w_i + G_i^{j\alpha\beta}(w,\partial w)\partial_{\alpha\beta} w_j + c_i^2 w_i = F_i(w,\partial w), \\ & w_i|_{\mathcal{H}_{B+1}} = w_{i0}, \quad \partial_t w_i|_{\mathcal{H}_{B+1}} = w_{i1} \end{aligned} \tag{11.2.4}$$

under the condition (on the initial slice)

$$\sum_i \sum_{|\beta|\leqslant l+1} \|\partial^\beta w_i\|_{L^2(\mathcal{H}_{B+1})} + \sum_i \sum_{|\beta|\leqslant l} \|\partial^\beta \partial_t w_i\|_{L^2(\mathcal{H}_{B+1})} \leqslant \epsilon \tag{11.2.5}$$

admits a unique local-in-time solution, which satisfies

$$\sum_i \sum_{|\beta|\leqslant l+1} \|\partial^\beta w_i\|_{L^2(\mathcal{H}_s)} \leqslant A\epsilon \tag{11.2.6}$$

in the time interval $[B+1, T(\epsilon)+B+1]$ and for soe constant $A > 0$. Furthermore, if T^* denotes the supremum of all such times $T(\epsilon)$ (for ϵ fixed), then either $T^* = +\infty$ or else

$$\lim_{\substack{s\to T^* \\ s<T^*}} \sum_i \sum_{\alpha\leqslant l+1} \|\partial^\alpha w_i\|_{L^2(\mathcal{H}_s)} = +\infty. \tag{11.2.7}$$

Bibliography

S. ALINHAC, The null condition for quasilinear wave equations in two space dimensions, II, American J. Math. 123 (2000), 1071–1101.

S. ALINHAC, The null condition for quasilinear wave equations in two space dimensions, I Invent. Math. 145 (2001), 597–618.

S. ALINHAC, An example of blowup at infinity for a quasilinear wave equation, Astérisque 284 (2003), 1–91.

S. ALINHAC, Remarks on energy inequalities for wave and Maxwell equations on a curved background, Math. Ann. 329 (2004), 707–722.

S. ALINHAC, Semilinear hyperbolic systems with blowup at infinity, Indiana Univ. Math. J. 55 (2006), 1209–1232.

S. ASAKURA, Existence of a global solution to a semi-linear wave equation with slowly decreasing initial data in three space dimensions, Comm. Partial Differential Equations 11 (1986), 1459–1487.

A. BACHELOT, Problème de Cauchy global pour des systèmes de Dirac-Klein-Gordon, Ann. Inst. Henri Poincaré 48 (1988), 387–422.

A. BACHELOT, Asymptotic completeness for the Klein-Gordon equation on the Schwarzschild metric, Ann. Inst. Henri Poincaré: Phys. Théor. 61 (1994), 411–441.

A. BACHELOT, The Klein-Gordon equation in the anti-de Sitter cosmology, J. Math. Pures Appl. 96 (2011), 527–554.

H. BAHOURI AND J.-Y. CHEMIN, Équations d'ondes quasilinéaires et inégalités de Strichartz, Amer. J. of Math. 121 (1999), 1337–1377.

F. BERNICOT AND P. GERMAIN, Bilinear dispersive estimates via space-time resonances: dimensions two and three, Arch. Ration. Mech. Anal. 214 (2014), 617–669.

T. CANDY, Bilinear estimates and applications to global well-posedness for the Dirac-Klein-Gordon equation on $\mathbb{R}^{1+1}$, J. Hyperbolic Differ. Equ. 10 (2013), 1–35.

D. CHRISTODOULOU, Global solutions of nonlinear hyperbolic equations for small initial data, Comm. Pure Appl. Math. 39 (1986), 267–282.

J.-M. DELORT, Existence globale et comportement asymptotique pour l'équation de Klein-Gordon quasi-linéaire à données petites en dimension 1, Ann. Sci.

École Norm. Sup. 34 (2001), 1–61.

J.-M. Delort and D. Fang, Almost global existence for solutions of semilinear Klein-Gordon equations with small weakly decaying Cauchy data, Comm. Partial Differential Equations 25 (2000), 2119–2169.

J.-M. Delort, D. Fang, and R. Xue, Global existence of small solutions for quadratic quasilinear Klein-Gordon systems in two space dimensions, J. Funct. Anal. 211 (2004), 288–323.

J. Frauendiener, Numerical treatment of the hyperboloidal initial value problem for the vacuum Einstein equations. II. The evolution equations, Phys Rev D 58 (1998), 064003.

J. Frauendiener, Some aspects of the numerical treatment of the conformal field equations. Lect. Notes Phys., 604 (2002) 261-282.

J. Frauendiener, Conformal infinity, Living Rev. Relativity 7, 2004.

H. Friedrich, On the regular and the asymptotic characteristic initial value problem for Einstein's vacuum field equations, Proc. R. Soc. London Ser. A 375 (1981), 169–184.

H. Friedrich, Cauchy problems for the conformal vacuum field equations in general relativity, Commun. Math. Phys. 91 (1983), 445–472.

H. Friedrich, Conformal Einstein evolution, Lect. Notes Phys. 604 (2002), 1–50.

V. Georgiev, Global solution to the system of wave and Klein-Gordon equations, Math. Z. 203 (1990), 683–698.

V. Georgiev, Decay estimates for the Klein-Gordon equation, Comm. Partial Differential Equations 17 (1992), 1111–1139.

V. Georgiev and P. Popivanov, Global solution to the two-dimensional Klein-Gordon equation, Comm. Partial Differential Equations 16 (1991), 941–995.

P. Germain, Global existence for coupled Klein-Gordon equations with different speeds, Ann. Inst. Fourier (Grenoble) 61 (2011), 2463–2506.

P. Germain and N. Masmoudi, Global existence for the Euler-Maxwell system, Ann. Sci. Éc. Norm. Supér. 47 (2014), 469–503.

P. Germain, N. Masmoudi and J. Shatah, Global solutions for the gravity water waves equation in dimension 3, Ann. of Math. 175 (2012), 691–754.

L. Hörmander, $L^1 L^\infty$ estimates for the wave operator, in "Analyse Math ématique et Applications", Contributions en l'honneur de J. L. Lions-Gauthier-Villars, Paris, 1988, pp. 211–234.

L. Hörmander, *Lectures on nonlinear hyperbolic differential equations,* Springer Verlag, Berlin, 1997.

A. Hoshiga and H. Kubo, Global small amplitude solutions of nonlinear hyperbolic systems with a critical exponent under the null condition, SIAM J. Math. Anal. 31 (2000), 486–513.

F. John, Blow-up of solutions of nonlinear wave equations in three space dimensions, Manuscripta Math. 28 (1979), 235–268.

F. John, Blow-up of solutions for quasi-linear wave equations in three space dimensions, Comm. Pure Appl. Math. 34 (1981), 29–51.

F. John, Lower bounds for the life span of solutions of nonlinear wave equations in three space dimensions, Comm. Pure Appl. Math. 36 (1983), 1–35.

F. JOHN, Existence for large times of strict solutions of nonlinear wave equations in three space dimensions for small initial data, Comm. Pure Appl. Math. 40 (1987), 79–109.

S. KATAYAMA, Global existence for coupled systems of nonlinear wave and Klein-Gordon equations in three space dimensions, Math. Z. 270 (2012), 487–513.

S. KATAYAMA, Asymptotic pointwise behavior for systems of semilinear wave equations in three space dimensions, J. Hyperbolic Differ. Equ. 9 (2012), 263–323.

S. KATAYAMA, Asymptotic behavior for systems of nonlinear wave equations with multiple propagation speeds in three space dimensions, J. Differential Equations 255 (2013), 120–150.

S. KATAYAMA AND H. KUBO, The rate of convergence to the asymptotics for the wave equation in an exterior domain, Funkcial. Ekvac. 53 (2010), 331–358.

S. KATAYAMA AND H. KUBO, Lower bound of the lifespan of solutions to semilinear wave equations in an exterior domain, J. Hyperbolic Differ. Equ. 10 (2013), 199–234.

S. KATAYAMA, T. OZAWA, AND H. SUNAGAWA, A note on the null condition for quadratic nonlinear Klein-Gordon systems in two space dimensions, Comm. Pure Appl. Math. 65 (2012), 1285–1302.

S. KATAYAMA AND K. YOKOYAMA, Global small amplitude solutions to systems of nonlinear wave equations with multiple speeds, Osaka J. Math. 43 (2006), 283–326.

Y. KAWAHARA AND H. SUNAGAWA, Global small amplitude solutions for two-dimensional nonlinear Klein-Gordon systems in the presence of mass resonance, J. Differential Equations 251 (2011), 2549–2567.

S. KLAINERMAN, Global existence for nonlinear wave equations, Comm. Pure Appl. Math. 33 (1980), 43–101.

S. KLAINERMAN, Global existence of small amplitude solutions to nonlinear Klein-Gordon equations in four spacetime dimensions, Comm. Pure Appl. Math. 38 (1985), 631–641.

S. KLAINERMAN, The null condition and global existence to nonlinear wave equations, in "Nonlinear systems of partial differential equations in applied mathematics", Part 1, Santa Fe, N.M., 1984, Lectures in Appl. Math., Vol. 23, Amer. Math. Soc., Providence, RI, 1986, pp. 293–326.

S. KLAINERMAN, Remarks on the global Sobolev inequalities in the Minkowski space $\mathbb{R}^{n+1}$, Comm. Pure Appl. Math. 40 (1987), 111–117.

S. KLAINERMAN AND M. MACHEDON, Estimates for null forms and the spaces H^s_δ, Internat. Math. Res. Notices 17 (1996), 853-865.

S. KLAINERMAN AND S. SELBERG, Bilinear estimates and applications to nonlinear wave equations, Commun. Contemp. Math. 4 (2002), 223–295.

S. KLAINERMAN AND T.C. SIDERIS, On almost global existence for nonrelativistic wave equations in 3d, Comm. Pure Appl. Math. 49 (1996), 307–321.

K. KUBOTA AND K. YOKOYAMA, Global existence of classical solutions to systems of nonlinear wave equations with different speeds of propagation, Japan. J. Math. (N.S.) 27 (2001), 113–202.

D. LANNES, Space-time resonances [after Germain, Masmoudi, Shatah],

Séminaire Bourbaki, Vol. 2011/2012, Astérisque No. 352 (2013), 355–388.

H. LINDBLAD, On the lifespan of solutions of nonlinear wave equations with small initial data, Comm. Pure Appl. Math. 43 (1990), 445–472.

H. LINDBLAD, M. NAKAMURA, AND C.D. SOGGE, Remarks on global solutions for nonlinear wave equations under the standard null conditions, J. Differential Equations 254 (2013), 1396–1436.

H. LINDBLAD AND I. RODNIANSKI, Global existence for the Einstein vacuum equations in wave coordinates, Comm. Math. Phys. 256 (2005), 43–110.

J. LUK, The null condition and global existence for nonlinear wave equations on slowly rotating Kerr spacetimes, J. Eur. Math. Soc. (JEMS) 15 (2013), 1629–1700.

V. MONCRIEF AND O. RINNE, Regularity of the Einstein equations at future null infinity, Class. Quant. Grav. 26 (2009), 125010.

C.S. MORAWETZ AND W.A. STRAUSS, Decay and scattering of solutions of a nonlinear relativistic wave equation, Comm. Pure Appl. Math. 25 (1972), 1–31.

C.S. MORAWETZ, Decay for solutions of the exterior problem for the wave equation, Comm. Pure Appl. Math. 28 (1974), 229–264.

T. OZAWA, K. TSUTAYA, AND Y. TSUTSUMI, Normal form and global solutions for the Klein-Gordon-Zakharov equations, Ann. Inst. H. Poincaré Anal. Non Linéaire 12 (1995), 459–503.

T. OZAWA, K. TSUTAYA, AND Y. TSUTSUMI, Global existence and asymptotic behavior of solutions for the Klein-Gordon equations with quadratic nonlinearity in two space dimensions, Math. Z. 222 (1996), 341–362.

R. PENROSE, Asymptotic properties of fields and space-times, Phys. Rev. Lett. 10 (1963), 66–68.

G. PONCE AND T.C. SIDERIS, Local regularity of nonlinear wave equations in three space dimensions, Comm. Partial Differential Equations 18 (1993), 169–177.

F. PUSATERI AND J. SHATAH, Space-time resonances and the null condition for first-order systems of wave equations, Comm. Pure Appl. Math. 66 (2013), 1495–1540.

F. PUSATERI, Space-time resonances and the null condition for wave equations, Boll. Unione Mat. Ital. 6 (2013), 513–529.

O. RINNE AND V. MONCRIEF, Hyperboloidal Einstein-matter evolution and tails for scalar and Yang-Mills fields, Class. Quantum Grav. 30 (2013), 095009.

J. SHATAH, Normal forms and quadratic nonlinear Klein–Gordon equations, Comm. Pure Appl. Math. 38 (1985), 685–696.

J. SHATAH, Space-time resonances, Quart. Appl. Math. 68 (2010), 161–167.

T.C. SIDERIS, Decay estimates for the three space dimensional inhomogeneous Klein-Gordon equation and applications, Comm. Partial Differential Equations 14 (1989), 1421–1455.

T.C. SIDERIS, Non-resonance and global existence of prestressed nonlinear elastic waves, Ann. Math. 151 (2000), 849–874.

T.C. SIDERIS AND S.-Y. TU, Global existence for systems of nonlinear wave equations in 3D with multiple speeds, SIAM J. Math. Anal. 33 (2001),

477–488.

C.D. Sogge, *Lectures on nonlinear wave equations,* International Press, Boston, 2008.

E.M. Stein, *Singular integrals and differentiability properties of functions,* Princeton University Press, 1970.

W.A. Strauss, Decay and asymptotics for $\Box u = F(u)$, J. Funct. Anal. 2 (1968), 409–457.

W.A. Strauss, *Nonlinear wave equations,* CBMS 73, Amer. Math. Soc., Providence, 1989.

H. Sunagawa, On global small amplitude solutions to systems of cubic nonlinear Klein-Gordon equations with different mass terms in one space dimension, J. Differential Equations 192 (2003), 308–325.

H. Sunagawa, A note on the large time asymptotics for a system of Klein-Gordon equations, Hokkaido Math. J. 33 (2004), 457–472.

D. Tataru, Strichartz estimates for operators with nonsmooth coefficients and the nonlinear wave equation I, Amer. J. Math. 122 (2000), 349–376.

D. Tataru, Strichartz estimates for operators with nonsmooth coefficients and the nonlinear wave equation II, Amer. J. Math. 123 (2001), 385–423.

D. Tataru, Strichartz estimates for operators with nonsmooth coefficients and the nonlinear wave equation III, J. Amer. Math. Soc. 15 (2002), 419–442.

M.E. Taylor, *Partial differential equations III. Nonlinear equations,* Appl. Math. Sc., Vol. 117, Springer Verlag, New York, 2011.

Y. Tsutsumi, Stability of constant equilibrium for the Maxwell–Higgs equations, Funkcial. Ekvac. 46 (2003), 41–62.

Y. Tsutsumi, Global solutions for the Dirac-Proca equations with small initial data in 3 + 1 space time dimensions, J. Math. Anal. Appl. 278 (2003), 485–499.

K. Yokoyama, Global existence of classical solutions to systems of wave equations with critical nonlinearity in three space dimensions, J. Math. Soc. Japan 52 (2000), 609–632.

A. Zenginoglu, Hyperboloidal evolution with the Einstein equations, Class. Quantum Grav. 25 (2008), 195025.

A. Zenginoglu, Hyperboloidal layers for hyperbolic equations on unbounded domains, J. Comp. Physics 230 (2011), 2286–2302.

A. Zenginoglu and L.E. Kidder, Hyperboloidal evolution of test fields in three spatial dimensions, Phys. Rev. D81 (2010) 124010.

Printed in the United States
By Bookmasters